DARWIN'S CATHEDRAL

DARWIN'S CATHEDRAL

EVOLUTION, RELIGION, AND THE NATURE OF SOCIETY

DAVID SLOAN WILSON

THE UNIVERSITY OF CHICAGO PRESS

Chicago and London

David Sloan Wilson is professor of biology and anthropology at Binghamton University. He is the author of *The Natural Selection of Populations and Communities* and co-author of *Unto Others: The Evolution and Psychology of Unselfish Behavior.*

The University of Chicago Press, Chicago, 60637
The University of Chicago Press, Ltd., London
© 2002 by The University of Chicago
All rights reserved. Published 2002
Printed in the United States of America

11 10 09 08 07 06 05 04 03 02   5 4 3 2 1
ISBN (cloth): 0-226-90134-3

Library of Congress Cataloging-in-Publication Data

Wilson, David Sloan.
    Darwin's cathedral : evolution, religion, and the nature of society / David Sloan
Wilson.
        p. cm.
    Includes bibliographical references (p.).
    ISBN 0-226-90134-3 (cloth : alk. paper)
    1. Religion and sociology.   2. Group selection (Evolution)   I. Title.
BL60.W544 2002
306.6—dc21

                                                          2002017375

♾ The paper used in this publication meets the minimum requirements of the American National Standard for Information Sciences—Permanence of Paper for Printed Library Materials, ANSI Z39.48-1992.

CONTENTS

Introduction
Church as Organism   *1*

Chapter 1
The View from Evolutionary Biology   *5*

Chapter 2
The View from the Social Sciences   *47*

Chapter 3
Calvinism: An Argument from Design   *86*

Chapter 4
The Secular Utility of Religion: Historical Examples   *125*

Chapter 5
The Secular Utility of Religion: The Modern Literature   *161*

Chapter 6
Forgiveness as a Complex Adaptation   *189*

Chapter 7
Unifying Systems   *219*

Notes   *235*

Bibliography   *245*

Acknowledgments   *259*

Index   *261*

But now at least he understood his religion: its essence was the relation between man and his fellows.

—Isaac Bashevis Singer

INTRODUCTION

CHURCH AS ORGANISM

> True love means growth for the whole organism, whose members
> are all interdependent and serve each other. That is the outward
> form of the inner working of the Spirit, the organism of the Body
> governed by Christ. We see the same thing among the bees, who
> all work with equal zeal gathering honey.
>
> —Ehrenpreis [1650] 1978, 11

Religious believers often compare their communities to a single organism or even to a social insect colony. The passage quoted above is from the writings of the Hutterites, a Christian denomination that originated in Europe five centuries ago and that currently thrives in communal settlements scattered throughout northwestern North America. Beehives are pictured on the road signs of the Mormon-influenced state of Utah. Across the world in China and Japan, Zen Buddhist monasteries were often constructed to resemble a single human body (Collcutt 1981).

The purpose of this book is to treat the organismic concept of religious groups as a serious scientific hypothesis. Organisms are a product of natural selection. Through countless generations of variation and selection, they acquire properties that enable them to survive and reproduce in their environments. My purpose is to see if human groups in general, and religious groups in particular, qualify as organismic in this sense.

Science works best when a subject can be resolved into well-framed hypotheses that make different predictions about measurable aspects of the world. Unfortunately, many subjects have a long hard road to travel before they reach that exalted state. The organismic concept of society has had an especially difficult journey. To some it is so self-evident that it scarcely requires testing. To others it is so far-fetched that it deserves mockery more than serious consideration. When smart people disagree to this extent, it is often because they are speaking different languages. Indeed, when one studies the organismic concept of society across time and disciplinary boundaries, one encounters a Tower of Babel. I am therefore faced with a second task almost as challenging as my first. Before I can treat the organismic concept of religious groups as a serious scientific hypothesis, I must provide a translation manual that allows meaningful communication across diverse fields of thought.

My inquiry requires equal attention to biology and to the vast literature devoted to our own species. Evolutionary theory explains how social groups can be like individuals in the harmony and coordination of their parts. Testing evolutionary theory requires a detailed knowledge of organisms in relation to their environment. When the organisms are religious groups, this means tackling an enormous literature from fields as diverse as theology, history, anthropology, sociology, and psychology. No one who has confronted this literature can claim to have mastered it, but I have made a solid effort and I expect to be judged by professional standards.

Evolutionary theories of human behavior often provoke great skepticism and hostility, inside and outside the ivory tower. Surprisingly, my own interactions with skeptics of all stripes are usually cordial and productive. One reason is that I approach others as a student rather than as a teacher. It is I who have much to learn from them, and if they also learn from my evolutionary perspective, so much the better. Another reason is that I agree with many of the criticisms leveled against past evolutionary theories of human behavior. I do not believe that human nature is fundamentally selfish, such that genuine altruism and morality become illusions. I do not believe that human nature can be explained entirely in terms of genetic evolution, such that it was set in stone during the Stone Age. I regard human evolution as a rapid and ongoing process, made possible by mechanisms loosely described as cultural, which means that human nature will never be set in stone, for better or for worse. Above all, I do not share the hostility that many evolutionists have expressed toward religion, from Huxley (1863) to Dawkins (1998).

A word about the scope of this book and the audience for which it is intended. Understanding religion requires answers to questions that extend beyond religion. What is the nature of human society? Is it a collection of self-seeking individuals, or can it be regarded as an organism in its own right? These questions have been pondered by inquiring minds throughout the ages, with no more agreement today than 500 or 2,000 years ago. Perhaps such big questions have no answers, forever remaining matters of opinion. Call me audacious, but I believe that they do have answers, based on developments in evolutionary biology that are only a few decades old. This book has been written for anyone who has wondered about the organismic nature of human society, regardless of their academic training or interest in religion per se.

I also hope that this book will be read by religious believers, despite its resolutely scientific approach to religion. Spirituality is in part a feeling of being connected to something larger than oneself. Religion is in part a collection of beliefs and practices that honor spirituality.[1] A scientific theory that affirms these statements cannot be entirely hostile to religion. I frankly admire many features of religion, without denying the many horrors that have also been committed in its name. Indeed, the hypothesis presented in this book explains the mix of blessings and horrors associated with religion better than any other hypothesis, or so I claim.

Religion is sometimes defined as a belief in supernatural agents. However, other people regard this definition as shallow and incomplete. The Buddha refused to be associated with any gods. He merely claimed to be awake and to have found a path to enlightenment. I am aware that Buddhism as actually practiced is often chock full of gods, but the opinion of its founder is still relevant. If there is more to religion than belief in supernatural agents, then perhaps science is not as hostile to religion as it is often taken to be. One reason that I admire some aspects of religion is because I share some of its values. I have not attempted to hide this fact, and I hope that it has not intruded upon my science. Nor have I attempted to conceal my own basic optimism that the world can be a better place in the future than in the past or present—that there can be such a thing as a path to enlightenment. Being a scientist does not require becoming indifferent to human welfare.

So, I have written a book for readers from all backgrounds, inside and outside academia and religion. Because I must start with the basics, I have been unable to discuss certain advanced issues as much as my own closest colleagues might like. Professional journals are the forum for such discussions, and the endnotes provide a guide to this literature. At the same time, I have not "dumbed down" the material for a popular audience. Serious intellectual work, even at a basic level, is not like eating candy. My ideal reader takes the same interest and pleasure in mental exercise that so many people try to cultivate for physical exercise. Becoming knowledgeable about a sport is vastly more exciting than simply being told the score. Playing the sport is more exciting still. This book is very much about science in motion, which invites the reader to become knowledgeable and perhaps even to play. As for the result—the game isn't yet over!

Biologists frequently express a feeling of awe, bordering on religious reverence, toward the intricacies of nature; the cryptic insect that exactly resembles a leaf, the fish that glides effortlessly through the water, and the amazing physiological processes that allow organisms to defy the forces of entropy. The organismic concept of groups makes possible a similar sense of awe toward religion, even from a purely evolutionary perspective.

CHAPTER 1

THE VIEW FROM EVOLUTIONARY BIOLOGY

> There can be no doubt that a tribe including many members
> who, from possessing in a high degree the spirit of patriotism,
> fidelity, obedience, courage, and sympathy, were always ready to
> aid one another, and to sacrifice themselves for the common
> good would be victorious over most other tribes; and this would
> be natural selection.
>
> —Darwin 1871, 166

Religion is often used to explain purpose and order at all levels, from celestial bodies, to human society, to the actions of individual people and other creatures. Darwin showed how the properties inherent in words such as "purpose," "order," "adaptation," and "organism" can arise by the process of natural selection. However, the evolutionary concept of purpose and order is highly restrictive and may well apply to individuals but not to groups. The image of society as a single organism writ large, which has so often been taken for granted by religious and other nonevolutionary thinkers, must be questioned very seriously in the light of evolution.

To evaluate religious groups as organisms, we must begin with the more general question of whether any kind of group qualifies as an organism. This chapter will lay the groundwork by reviewing the history of thinking on groups in evolutionary biology, from Darwin to the present, with special reference to human evolution (see Sober and Wilson 1998 for a book-length account). It is a tumultuous and fascinating history. Although Darwin was characteristically clear-sighted, many of his successors uncritically assumed that adaptive societies can evolve as easily as adaptive individuals. The term "for the good of the group" was used as freely as "for the good of the individual." Then, starting in the 1960s, adaptation at the level of groups was rejected so strongly that the ensuing period could be called the age of individualism in evolutionary biology. Fortunately, science is not destined to be a frictionless pendulum that swings back and forth between extreme positions. A middle ground is becoming established in which groups are acknowledged to evolve into adaptive units, but only if special conditions are met. Ironically, in human groups it is often religion that provides the special conditions. Religion returns to center stage, not as a theological explanation of purpose and order, but as itself a product of evolution that enables groups to function as adaptive units—at least to a degree.

ADAPTATION AND FUNCTIONALIST THINKING

The basic concept of adaptation and the interpretation of groups as adaptive units existed far before Darwin's theory of evolution. The *Oxford English Dictionary* defines adaptation as "the action or process of fitting or suiting one thing to another." Human artifacts are often clearly adapted to perform a given function. A bow is designed to shoot an arrow. An arrow is designed to fly through the air and pierce its target. Likewise, the form and behavior of organisms are often clearly adapted to achieve certain goals. The coat of a polar bear is designed to keep it warm and to prevent it from being seen. The behavior of a polar bear stalking its prey is designed to get close enough to attack while avoiding detection.

Thinking about an object or an organism as if it has a purpose can be called functionalist thinking. Functionalist thinking can be highly effective when applied to things that actually have a purpose, but in other contexts it can be misleading. Wondering about the purpose of your neighbor's behavior can help you discern his intentions, but wondering about the purpose of the moon leads only to a folk tale. The reverse is true for nonfunctionalist thinking. A boulder rolling down a hill has no purpose, but merely a path which must be predicted to get out of the way. It would be disastrous to think of a boulder as like an attacking predator or an attacking predator as like a boulder. Functionalist and nonfunctionalist ways of thought are so different from each other, and so useful in some contexts but misleading in others, that they may actually have evolved as separate cognitive skills (Hauser and Carey 1998; Tomasello 1999).

Functionalist thinking has been applied to social groups throughout history. Plato compared the various classes of society to the organs of a single organism. Religious thought is rife with organismic allusions—as in Paul's description (1 Cor. 12) of the body of the church, united under the head of Christ. Similarly, the Anglican bishop Joseph Butler ([1726] 1950, 21) claimed that "it is as manifest that we were made for society and to promote the happiness of it, as that we were intended to take care of our own life and health and private good." The founders of the social sciences talked unabashedly about societal organisms, complete with group minds (Wegner 1986). However, although the metaphor of society as organism appears plausible in some cases, it is misleading in others. A well-organized attacking army seems more like a single predator than a passing boulder, but the class of beggars has no organ-like function in society, although it may be explained by functionalist thinking at a lower level, such as individual greed resulting in an unequal division of resources. Thus, social groups are a nebulous and heterogeneous category with respect to the concept of adaptation.

The Fundamental Problem of Social Life

Darwin provided the first successful scientific theory of adaptations. Evolution explains adaptive design on the basis of three principles: phenotypic variation, heritability, and fitness consequences. A phenotypic trait is anything that can be observed or measured. Individuals in a population are seldom identical and usually vary in their phenotypic traits. Furthermore, offspring frequently resemble their parents, sometimes because of shared genes but also because of other factors such as cultural transmission. It is important to think of heritability as a correlation between parents and offspring, caused by any mechanism. This definition will enable us to go beyond genes in our analysis of human evolution. Finally, the fitness of individuals—their propensity to survive and reproduce in their environment—often depends on their phenotypic traits. Taken together, the three principles lead to a seemingly inevitable outcome—a tendency for fitness-enhancing phenotypic traits to increase in frequency over multiple generations. Darwin's theory is so simple that it can be explained in a single paragraph, but its implications are so profound that the study of life was transformed, enabling the great geneticist Theodosius Dobzhansky (1973) to say: "Nothing in biology makes sense except in the light of evolution."

The fact that adaptation is defined in terms of survival and reproduction places limits on the kinds of adaptation that can evolve. To appreciate the limitations, let's first consider the evolution of an individual-level adaptation, such as cryptic coloration. Imagine a population of moths that vary in the degree to which they match their background. Every generation, the most conspicuous moths are detected and eaten by predators while the most cryptic moths survive and reproduce. If offspring resemble their parents, then the average moth will become more cryptic with every generation. Anyone who has beheld an insect that looks exactly like a leaf, right down to the veins and simulated herbivore damage, cannot fail to be impressed by the ability of natural selection to produce breathtaking adaptations at the individual level.

Now consider the same process for a group-level adaptation, such as members of a group warning each other about approaching predators. Imagine a flock of birds that vary in their tendency to scan the horizon for predators and to utter a call when one is spotted. The most vigilant individuals will not necessarily survive and reproduce better than the least vigilant. If scanning the horizon detracts from feeding, the most vigilant birds will gather less food than their more oblivious neighbors. If uttering a cry attracts the attention of the predator, then sentinels place themselves at risk by warning others. Birds that do not scan the horizon and that remain silent when they see a predator may well survive and reproduce better than their vigilant neighbors.

These two examples show that the evolutionary concept of adaptation does not always conform to the intuitive concept, especially at the group level. It is easy to imagine a bird flock as an adaptive unit and to use functionalist thinking to predict its properties. We would expect members of the flock to adopt the creed "all for one and one for all." We might expect sentries to be posted at all times to detect predators at the earliest possible moment and to relay the information to feeding members of the flock. Unfortunately, individuals who display these prosocial behaviors do not necessarily survive and reproduce better than those who enjoy the benefits without sharing the costs. Since Darwin's theory relies entirely on differences in survival and reproduction, it appears unable to explain groups as adaptive units. This can be called the fundamental problem of social life. Groups function best when their members provide benefits for each other, but it is difficult to convert this kind of social organization into the currency of biological fitness.

Now we can begin to see why the concept of religious groups as adaptive units does not emerge automatically from evolutionary theory. On the basis of what we have considered so far, the theory has difficulty explaining any kind of group as an adaptive unit, including those that might be found in our own species.

Darwin's Solution to the Fundamental Problem

Darwin was aware of the fundamental problem of social life and proposed a solution. Suppose there is not just one flock of birds but many flocks. Furthermore, suppose that the flocks vary in their proportion of callers. Even if a caller does not have a fitness advantage within its own flock, groups of callers will be more successful than groups of noncallers. In the following famous passage from *The Descent of Man*, Darwin (1871, 166) used this reasoning to explain the evolution of human moral virtues that appear designed to promote group welfare:

> It must not be forgotten that although a high standard of morality gives but a slight or no advantage to each individual man and his children over the other men of the same tribe, yet that an increase in the number of well-endowed men and advancement in the standard of morality will certainly give an immense advantage to one tribe over another. There can be no doubt that a tribe including many members who, from possessing in a high degree the spirit of patriotism, fidelity, obedience, courage, and sympathy, were always ready to aid one another, and to sacrifice themselves for the common good would be victorious over most other tribes; and this would be natural selection. At all times throughout the world tribes have supplanted other tribes; and as morality is one important element in their success, the standard of morality and the number of well-endowed men will thus everywhere tend to rise and increase.

Darwin was proposing that the three ingredients of natural selection—phenotypic variation, heritability, and fitness consequences—can exist at the level of groups. There can be a population of groups (many tribes of humans, many flocks of birds) that vary in their phenotypic properties (standard of morality, warning cries), with consequences for survival and reproduction (intertribal warfare, avoiding predators). If current groups resemble the previous groups from which they were derived, then groups can evolve into adaptive units in just the same way that individuals evolve into adaptive units.

Darwin's solution to the fundamental problem of social life is elegant and perhaps even obvious in retrospect. After all, if adaptations evolve by differential survival and reproduction, it makes sense that group-level adaptations evolve by the differential survival and reproduction of groups. However, Darwin's solution has two limitations that must always be kept in mind. First, just because groups can evolve into adaptive units doesn't mean that they do. The days of axiomatically thinking of groups as adaptive units are gone forever. Special conditions are required that may or may not be satisfied in the real world. Opposing forces exist that may or may not be overcome. In the case of our birds, group selection favors vigilant callers but selection within groups favors birds that stuff their crops and think only of saving their own feathers when a predator appears on the horizon. If we wish to explain bird flocks as adaptive units, not only must we demonstrate a process of among-group selection, but we also must show that it operates more strongly than the opposing process of within-group selection. The term multilevel selection expresses the possibility that natural selection can operate at more than one level of the biological hierarchy.

Second, even when groups do evolve into adaptive units, often they are adapted to behave aggressively toward other groups. In Darwin's scenario, the moral virtues are practiced among members of a tribe and are directed against other tribes. Group selection does not eliminate conflict but rather elevates it up the biological hierarchy, from among individuals within groups to among groups within a larger population. The most that group selection can do is produce groups that are like organisms in the harmony and coordination of their parts. We already know about the competitive and predatory interactions that take place among individual organisms in ecological communities, and the same can be expected of well-adapted groups. This might be a disappointment for those searching for a universal morality that transcends group boundaries, but it follows directly from the organismic concept of groups. I do not mean to imply that the search for a universal morality is hopeless, only that it does not follow automatically from group selection theory. Religions are well known for their in-group morality and out-group hostility, so we will return to this theme repeatedly in future chapters.

Looking forward, we can anticipate that evolutionary theory will turn the study of religion into quite a complex subject. It is possible to imagine religious groups as adaptive units, but this outcome is by no means obvious or inevitable. A major alternative hypothesis is that some features of religion are a product of within-group selection, benefiting some individuals at the expense of others within the same religious group. In addition, a host of nonfunctional explanations are possible, since there is more to evolution than natural selection (Gould and Lewontin 1979; Williams 1996). The likelihood of these possibilities depends on many factors, including the balance between levels of selection. To make matters even more complex, genetic and cultural evolution are both multilevel processes that interact with each other. This book is about evolution but it is not restricted to genetic evolution. At all times throughout the world (to paraphrase Darwin) religious systems have arisen in profusion, competing against each other and against nonreligious social organizations. Differences among religions are culturally based, but that does not prevent religious groups from succeeding or failing on the basis of their properties and for these properties to be transmitted with modification to descendant groups. At a different temporal scale, the human mind is a product of genetic evolution over thousands of generations during which people were subdivided into small groups of hunter-gatherers. Our genetically innate psychology might therefore reflect the influence of both within- and among-group selection, regardless of the kinds of groups in which we participate today (Barkow, Cosmides, and Tooby 1992; Boehm 1999). Evaluating these possibilities and relating them to the nature of modern religious groups will be a major undertaking requiring many scholar-decades of work.

EVOLUTIONARY THEORY'S WRONG TURN

I have portrayed group selection as a process that can occur but which also must contend against forces that pull in other directions. In the 1960s a consensus emerged that group selection is such a weak force that it can be ignored for most purposes (Williams 1966). The consensus held that even though it is theoretically possible for groups to evolve into adaptive units, it almost never happens in the real world. Konner (1999, 30–31) describes this period in the history of evolutionary biology:

> Current intrusions of Darwin's theory into our awareness stem from the mid-1960s, when the British geneticist W. D. Hamilton proposed a solution to the problem of altruism. For traditional social scientists who see societies as functioning organisms, the existence of altruism does not pose a problem. In this view, without altruism societies would not work; groups that lacked it would not survive.
>
> But this is no comfort to strict Darwinians, who see natural selection as operating at the level of individuals, even to the extent of disrupting the cohesiveness of societies. In their view, natural selection should have long since erased altruism. Hamilton's solution was that evolution selects for altruism if it is directed at relatives in proportion to their relatedness, for then the altruist's kin are more likely to survive to pass on the contrib-

uting genes. . . . Reciprocal altruism, proposed by Robert Trivers in the early 1970s, was a you-scratch-my-back-and-later-I'll scratch-yours model. Like kin selection, it required no real genetic generosity, only delayed self-interest. With these ideas, biologists seemed to have little further need for the metaphor of society as an organism.

Konner appreciates the diversity of perspectives across time and disciplines that I have also emphasized. Although many social scientists take the or-ganismic concept of society for granted, evolutionary biologists in the 1960s rejected group selection so strongly that it became heretical to think of "society as an organism"—to use Konner's words—for humans or any other species. Individuals are the organisms and society is merely a conve-nient word for what individuals do to each other in the course of maximiz-ing their own fitness. The illusion of adaptation at the group level can be explained in terms of individuals increasing the fitness of their genes in the bodies of others, reciprocal exchange, or even more self-serving bene-fits such as downright deception and exploitation.

The rejection of group selection was hailed by evolutionary biologists as a major event. Alexander (1987, 3) even called it the greatest intellectual revolution of the twentieth century. It is true that the early group selection literature was an easy target for criticism. When a biologist explained a given behavior as for the good of the group or the species, it was usually a naive expression of group-level functionalism rather than a principled argument. However, the wholesale rejection of group selection was itself a wrong turn from which the field is only starting to recover. I have written extensively on this topic elsewhere, including my book-length account with Elliott Sober (Sober and Wilson 1998; Wilson 1998a, 1999a, 2000). Here I will provide a brief summary of evolutionary theory's wrong turn and why it needs to be put behind us.[1]

How to see group selection

To give the critics of group selection their due, it is perfectly possible for a behavior that seems altruistic to be individually selfish upon closer inspection.[2] Returning to our bird example, suppose that uttering a call does not increase the risk of being attacked by an approaching predator. On the contrary, calling advertises to the predator that it has been spotted and that a less vigilant member of the group should be targeted. If these are the facts of the matter, then the evolution of calling behavior could be explained entirely by within-group selection. Calling individuals survive and reproduce better than noncalling individuals in the same group. The function of the adaptation is not to warn other members of the group but to communicate with the predator in a way that actually endangers other members of the group. We would be right to reject group selection in this case.

But now suppose that our original story is correct and calls are used to warn other members of the flock at the caller's expense. Calling is selectively disadvantageous within groups and evolves only because groups of callers fare better than groups of noncallers. A subtle shift in perspective can make calling appear individually selfish, even though it evolves by group selection. To pick an extreme example, imagine a flock of birds with one caller and nine noncallers. Everyone has a low fitness in this flock because only one bird is looking out for predators; however, this bird has the lowest fitness of all. Let us say that the chance of surviving predators is 50 percent for the noncallers and 25 percent for the caller. A second flock of birds has nine callers and one noncaller. Everyone has a high fitness in this flock because so many members are looking out for predators; however, the shirking noncaller has the highest fitness of all. Let us say that the chance of surviving predators is 100 percent for the noncaller and 75 percent for the callers. When we compare the fitness of callers and noncallers within each group, we see that callers are the losers in both cases. However, the group with more callers fares better than the group with fewer callers. This is the classic group selection scenario that began with Darwin. Now for the subtle shift in perspective: Let's calculate the average survival of callers and noncallers across the groups. One noncaller has a survival probability of 100 percent and nine have a survival probability of 50 percent for an average of 55 percent. One caller has a survival probability of 25 percent and nine have a survival probability of 75 percent for an average of 70 percent. The average caller is more fit than the average noncaller, so why not say that calling evolves by individual selection? Like a magician's trick, the need to invoke group selection appears to vanish! Of course, the disappearance is just an illusion. The need for multiple groups and variation among groups is absolutely essential for the calling behavior to evolve.

It follows that a certain procedure is required to see group selection. First, we must identify the relevant groups, a point to which I will return below. Second, we must compare the fitnesses of individuals within groups. Third, we must compare the fitnesses of groups in the total population. Finally we must combine these effects to determine the net result of what evolves. Employing this procedure for our bird flock example, the groups are flocks, callers are less fit than noncallers within flocks, but flocks with more callers are more fit than flocks with fewer callers. When the variation among groups is as extreme as in my example (one caller in the first group and nine callers in the second group) group selection is by far the strongest evolutionary force and the calling behavior evolves despite its selective disadvantage within groups.[3] However, all of this clarity is lost when we average the fitness of individuals across groups. In this case we correctly conclude that the calling behavior evolves, but we are unable to say whether it evolves on the strength of a fitness advantage within groups or between groups—the very distinction required to determine if calling evolves by group selection! If we define "individual selection" in terms of fitness averaged across groups rather than fitness within single groups, we have defined group selection out of existence, making "individual selection" a vacuous term for "whatever evolves." Elliott Sober and I call the practice of first subsuming group selection into the definition of individual selection, and then using this expanded definition to argue against group selection, "the averaging fallacy."[4]

It turns out that the rejection of group selection in the 1960s was based largely on the averaging fallacy. The verdict seemed to be that within-group selection is invariably stronger than between-group selection, but the theories that replaced group selection also assumed the existence of groups. How could they fail to, since social interactions almost invariably take place in groups that are small compared to the total population? To proceed further, we need to confront the crucial question of how groups are defined in evolutionary models of social behavior.

How to define groups

The procedure for seeing group selection requires an unambiguous definition of groups. At first this might seem like a hopeless enterprise. The early biologists who thought uncritically about "the good of the group" saw adaptive groupings everywhere: from bird flocks, ant colonies, and fish schools at one end of the spectrum to whole species and ecosystems at the other end. The groups considered by modern multilevel selectionists are only slightly less diverse. Nevertheless, this diversity only exists when we consider a diversity of phenotypic traits. When we consider the evolution of a single trait, there is much less ambiguity about what constitutes a group.

Let's begin with our trusty bird example before deriving the general rule. Why did bird flocks seem like such appropriate groups for the evolution of warning calls? Because it was assumed as part of the example that all members of a flock are at risk from the approaching predator and hear the cry of any member of the flock. Uttering a cry alters fitness within the flock and has no effect on other flocks. If it turned out that the predator targets several flocks at once and that a warning cry is heard by all the flocks, we would need to change our definition of groups. Similarly, if it turned out that a cry is only heard by one's nearest neighbor rather than the whole flock, we might need to change our definition of groups yet again. The reason that the definition of groups is so closely tied to the details of the trait is because we are trying to predict the evolution of the trait. When the trait is a nonsocial behavior that alters the fitness of the individual alone, we needn't concern ourselves with groups. But when the trait is a social behavior, the fitness of an individual is determined by its own trait and the traits of the individuals with whom it interacts. These individuals constitute the group, which must be identified accurately to calculate the fitnesses that determine the outcome of evolution.[5]

It follows that groups must be defined separately for each and every trait. Suppose that we decide to study resource conservation in our birds. Birds that eat moderately have fewer offspring than birds in the same group that stuff their crops, but groups of moderate birds persist while groups of gluttonous birds overexploit their resources and go extinct (as proposed by Wynne-Edwards 1962). This is the same problem of altruism and selfishness that we encountered for warning cries, but here we need to find the appropriate groups for resource conservation. Bird flocks may be appropriate if they live on exclusive territories but not if many flocks share the same resources. If the birds live on an archipelago of islands, then perhaps a single island is the appropriate group for this particular trait. However, group selection might be less effective at the scale of islands than at the scale of bird flocks. If so, then the birds might behave altruistically with respect to warning cries but not with respect to feeding.

I coined the term "trait-group" to emphasize the intimate relationship between traits and groups in multilevel selection theory (Wilson 1975). However, this term merely recognizes something that has always been implicit in the definition of groups, inside and outside of biology. My bowling group is the people with whom I bowl, my study group is the people with whom I study, my platoon is the group of people with whom I fight, my nation is the group of people who share the same set of laws, my church is the group of people with whom I worship. All of these groups are defined in terms of the individuals who interact with respect to a given activity. There is an infinite variety of groups, but only when we consider an infinite variety of activities. For any particular activity, there is a single appropriate grouping. We all use the concept of trait-groups in our everyday lives without thinking about it.

Returning to biology, all theories of social behavior must identify the appropriate grouping to explain the evolution of a given behavior. Let's say that the behavior involves fighting over a resource. We imagine one type of individual who shares amicably and another type who fights to take it all. If the individuals always meet in pairs, then pairs are the appropriate groups. If the individuals meet in triads or in free-for-alls of one hundred, then those are the appropriate groups. The groups are decided by the biology of the organism, not the whim of the biologist. Evolutionary game theory, one of the theoretical frameworks that was developed as an alternative to group selection, is very careful in this regard (Maynard Smith 1982; Dugatkin 1998; Skyrms 1996). Its formal name is N-person evolutionary game theory, where N is the number of individuals who socially interact, thereby influencing each other's fitness. A game theory model of fighting, alarm calls, or resource conservation must identify the same groupings as a group selection model of fighting, alarm calls, or resource conservation. The same is true for all other theories of social behavior that were developed as alternatives to group selection, such as inclusive fitness theory (Hamilton 1964, 1975) and selfish gene theory (Williams 1966; Dawkins 1976).

The nonarbitrary definition of groups and the need for all theories to converge upon the same definition for any particular trait allows the averaging fallacy to be revealed to its fullest extent. Take any evolutionary theory of social behavior, including those that were developed as alternatives to group selection, and find the groups that are identified within their own frameworks. Employ the procedure outlined above for seeing group selection, and you will find it. Most of the behaviors that have been rendered as only "apparently" altruistic but "really" selfish during the age of individualism are selectively disadvantageous within groups and evolve only by increasing the fitness of groups, relative to other groups, exactly as Darwin proposed.[6]

It should be clear by now that the revival of multilevel selection theory involves more than acknowledging one or two examples of group selection. Once the averaging fallacy is avoided, group selection is required to explain many traits whose evolution is already well documented and accepted. However, remember that my own account of group selection was cautious. Just because it happens doesn't mean that it happens all the time or invariably prevails against within-group selection. There are plenty of cases in which group selection remains a meager evolutionary force, even after we see it clearly. The importance of group selection in human genetic and cultural evolution remains to be determined. Above all, we do not want to return to the days when groups were axiomatically assumed to function as adaptive units. The point is to achieve a middle ground in which the importance of the various levels of natural selection are examined on a case-by-case basis, using a procedure that allows group selection to be seen where it exists.[7] The belief that group selection can be categorically rejected belongs on the rubbish heap of history, alongside the earlier belief that groups always function as adaptive units.

MAJOR TRANSITIONS OF LIFE

The trait-group concept conflicts with the image of an organism as a unit that is adaptive with respect to many traits. After all, an individual organism like a bird eats as a unit, flies as a unit, fights as a unit, and so on. Some animal groups such as social insect colonies are integrated with respect to many traits. Similarly, some human groups organize the lives of their members from cradle to grave. In many other cases, however, groups are adaptive only with respect to one or a few traits. When I use the term "organismic" in connection with groups, it will be synonymous with "adaptive at the group level" and will refer to particular traits and the appropriate groupings for those traits, while remaining agnostic about other traits and groupings.[8] The fact that people often participate in many groupings whose adaptedness must be evaluated on a case-by-case basis will become clear in subsequent chapters.

When group selection was rejected in the 1960s it was believed that evolution takes place entirely by mutational change. Since then, it has become increasingly certain that evolution also takes place along a different pathway: by social groups becoming so functionally integrated that they become higher-level organisms in their own right. One of the first to propose this radical new theory was Lynn Margulis (1970), who claimed that eukaryotic cells—the nucleated cells of all organisms other than bacteria—are actually symbiotic communities of bacteria whose members led a more autonomous existence in the distant past. Now it appears likely that similar transitions, from groups *of* organisms to groups *as* organisms, have occurred throughout the history of life, right down to the origin of life itself as social groups of cooperating molecular reactions (Maynard Smith and Szathmary 1995; Michod 1999a).

These discoveries are deliciously ironic. Thinking of social groups as like organisms has been out of fashion for thirty-five years, and now it turns out that organisms are themselves social groups! Moreover, each transition requires group-level selection, exactly as Darwin proposed. For example, a bacterial cell can be regarded as a social group of genes that coordinate their activities for their collective benefit. However, this group can be exploited by genes that use the resources of the cell to replicate themselves rather than by making products that contribute to the common good. Such "selfish genes" would fare better than "solid citizen genes" within the same cell, but cells with "selfish genes" would fare more poorly than cells with "solid citizen genes." As an aside, my use of the term "selfish gene" in the previous sentence is not the same as its meaning in selfish gene theory (Dawkins 1976). In my case, I refer to genes that gain at the expense of other genes in the same cell; in contrast, selfish gene theory would classify either gene as selfish depending on which evolves, replacing the other gene in the total population (the averaging fallacy).[9] In any case, the problem of "solid citizen" vs. "selfish" genes within cells is identical to the problem of calling vs. noncalling birds within flocks and moral vs. immoral people within tribes that led Darwin to propose his theory of group selection.

Viewing single organisms as highly integrated social groups has vastly expanded the scope and importance of multilevel selection theory. As Robert Trivers once remarked in a lecture, those interested in the evolution of social behavior have always appreciated the need to understand genetics, but who would have guessed thirty years ago that geneticists would need to understand the evolution of social behavior? There is also a return benefit, for understanding the mechanisms that allow organisms to become so integrated can help to identify similar mechanisms in social groups that might previously have been overlooked. In particular, sociobiologists have been fascinated, even mesmerized, by the problem of explaining altruistic traits that benefit the group at great cost to the individual. As we have seen, group selection can produce altruistic traits, but it must be exceptionally strong to oppose the strong selective disadvantage of altruism within groups. In contrast, the mechanisms that allow organisms to function as adaptive units do not appear very altruistic. Returning to the "selfish gene" that replicates rather than contributing to the common good of the cell, this problem can be solved by linking all the genes together into a chromosome that replicates as a unit. By eliminating the possibility of differential replication within the cell, chromosomes concentrate the process of natural selection at the among-cell level, neatly solving the fundamental problem of social life. But the genes responsible for the evolution of the chromosome do not appear self-sacrificial. Instead, they appear to benefit the group, of which they are a part, at no special cost to themselves.

Social control, rather than highly self-sacrificial altruism, appears to solve the fundamental problem of social life at the individual level. An entire lexicon of words describing social control in human life has been borrowed to describe genetic and developmental interactions; "sheriff" genes, "parliaments" of genes, "rules of fairness," and so on. The laws of genetics and development, which originally referred merely to general patterns, have acquired an eerie resemblance to the other meaning of the word law—a social contract enforced by punishment.[10]

What works for individuals can also work for social groups. In their drive to explain highly self-sacrificial altruism, sociobiologists have tended to ignore an even more important question: Does benefiting the group require overt altruism on the part of individuals? If not, then group selection can favor mechanisms that organize groups into adaptive units without strong selection against these mechanisms within groups.

I use the word "overt" because a close look at social control mechanisms shows that they differ from altruism only in degree and not in kind. Returning to our bird example, suppose we discover that warning calls are indeed risky and help others at the expense of the caller. If they were performed voluntarily they would qualify as altruistic, with all the self-sacrifice implied by the word. Then we discover that they are not performed voluntarily because birds that fail to call are severely punished by other birds. Calling no longer qualifies as altruistic, but we still must explain the evolution of the punishing behavior that makes calling selfish. Punishers cause birds to issue warning calls that help everyone in the group, including free-riders who do not share the cost of enforcement. We have not solved the problem of altruism but merely moved it from the calling behavior to the punishment behavior. Economists call this a second-order public goods problem: causing another to perform a public good is itself a public good (Heckathorn 1990, 1993). There is, however, an important difference between the two kinds of altruism. The individual cost of enforcement can be much lower than the individual cost of issuing a warning cry. Social control can be regarded as a form of low-cost altruism that evolves to promote behaviors that would qualify as high-cost altruism if they were performed voluntarily. Elliott Sober and I call this "the amplification of altruism" (Sober and Wilson 1998, chap. 4). In general, social control mechanisms do not alter the basic conclusion that group-level adaptations require a corresponding process of group selection. Instead, they partially relax the trade-off between group benefit and individual self-sacrifice, allowing among-group selection to act without strong counteracting within-group selection.

The concept of organisms as social groups has transformed our understanding of multilevel selection in several ways. First, never again can it be said that higher-level selection is always weak compared to lower-level selection. Single organisms such as you and I are shining contradictions of that statement. Second, higher-level selection has always appeared unlikely because it has been linked with self-sacrificial altruism. Social control mechanisms cut this Gordian knot by partially relaxing the trade-off between group benefit and individual cost. Social control mechanisms are obviously relevant to religious groups, which are based on much more than voluntary altruism. Third, it is inconceivable that higher-level selection stops at the level currently known as individual organisms. Selection at the level of social groups is likely to be an important, if not a dominating, evolutionary force in thousands of species. In some cases such as the social insects, the groups are so thoroughly integrated that they deserve to be called organisms in their own right, as Wheeler (1928) suggested long ago and as modern social insect biologists such as Seeley (1995) increasingly acknowledge.

Against this background, the organismic concept of human groups receives new life. Thirty years ago, evolutionary biologists would have dismissed the Hutterites' comparison of their communities to bodies and beehives as the worst kind of naive group selectionism. Now it is a vivid dot on the scientific radar screen.

HUMAN GROUPS AS ADAPTIVE UNITS

So far I have discussed basic evolutionary principles that apply to all organisms. Now it is time to focus on our own species. We evolved in small groups that are roughly approximated by modern hunter-gatherer societies, which, though fast disappearing, still dot the surface of the globe. As Konner states in the passage quoted above, evolutionary biologists have tended to regard ancestral human groups as mere collections of self-interested individuals, exhibiting nepotism and niceness toward those who can return the favor but by no means qualifying as societal organisms.[11] Multilevel selection theory makes it appear more likely that ancestral human groups were potent units of selection (Boehm 1999; Sober and Wilson 1998).

First, some empirical facts. Anthropologists don't agree on much, but they appear to agree that modern hunter-gatherer societies around the world are remarkably egalitarian. The most impressive fact is that meat is usually scrupulously shared. The successful hunter and his immediate family get no more than the rest of the band. The most careful studies have weighed the meat on portable scales as it is divided into portions (Kaplan and Hill 1985a, b; Kaplan, Hill, and Hurtado 1984). Even when averaged over a period of weeks, there was no bias in favor of the actual procurers of the meat. Gathered items are shared less fully, but only in comparison to meat; the same study that reported 100 percent sharing of meat reported approximately 50 percent sharing of gathered items. If a number of people are gathering, it may make little sense to put the harvest together just to divide it again, so the sharing of gathered items must be evaluated differently than the sharing of meat.

Hunter-gatherer egalitarianism extends beyond food to social relationships. The request "take me to your leader" would be met with incomprehension, or perhaps ridicule, by a hunter-gatherer. There are no leaders other than those who have earned the respect of their peers by being models of good conduct, and who can only advise and not dictate. When the British anthropologist E. E. Evans-Pritchard attempted to identify leaders among the Nuer (a pastoralist rather than a hunter-gatherer society but similar with respect to egalitarianism), all he could find was someone called the leopard-skin chief who turned out to be a specialist in conflict resolution, about whom more will be said in chapters 2 and 6.

Hunter-gatherers are egalitarian, not because they lack selfish impulses but because selfish impulses are effectively controlled by other members of the group. This form of guarded egalitarianism has been called "reverse dominance" by anthropologist Chris Boehm (1993, 1999; see also Knauft 1991). In many animal groups, the strongest individuals are usually able to dominate their rivals, taking a disproportionate share of the resources. This is within-group selection pure and simple. In human hunter-gatherer groups, an individual who attempts to dominate others is likely to encounter the combined resistance of the rest of the group. In most cases even the strongest individual is no match for the collective, so self-serving acts are effectively curtailed. Boehm's survey of hunter-gatherer societies includes many examples of reverse domination, ranging in intensity from gossip, to ridicule, to ostracism, to assassination.

Earlier theories of hunter-gatherer egalitarianism focused on ecological conditions such as dispersed and unpredictable resources. In contrast, Boehm explains egalitarianism in terms of social norms, a shared understanding of do's and don'ts that are enforced by rewards and punishments. A hunter-gatherer society is above all a moral community with a strong sense of right and wrong that organizes the practices of the group. The specific practices regarded as right and wrong might vary across groups, but in general "right" coincides with group welfare and "wrong" coincides with self-serving acts at the expense of other members of the group.

The concept of human groups as moral communities shows how much has been missed by kin selection theory, which predicts that prosocial behaviors should be directed primarily toward genetic relatives.[12] Kin selection models assume that behavioral similarity is proportional to genetic similarity; for example, the only way to get a behaviorally uniform group is for it to be genetically uniform. In moral communities, social norms can create a degree of behavioral uniformity within groups and differences among groups that could never be predicted from their genetic structure and which is highly favorable for among-group selection.

The concept of human groups as moral communities also shows how much has been missed by the concept of reciprocal altruism, which predicts that prosocial behavior should be directed primarily toward those who will return the favor. Consider a group whose members believe that it is right to help others in proportion to need rather than the likelihood of return gains. Individuals who abide by the norm are rewarded, those who violate the norm are punished, and the group (let us say) prospers compared to groups whose members restrict helping to those who will return the favor. This is a plausible model, and there is no reason to think of the rewards and punishments supporting the norm of giving according to need as themselves a form of reciprocity. If I shun narrow reciprocators and favor those who give according to need, I am a second-order public good provider, not someone who is maximizing my own return benefits within my group.

On the other hand, the concept of human groups as moral communities fits nicely with the emerging paradigm of major transitions, in which groups become unified by a regulatory apparatus that promotes the welfare of the group as a whole without necessarily requiring extreme self-sacrifice of its members. An example will show how a real hunter-gatherer society accomplishes this, using mechanisms that border upon religion.

The Chewong are a tribe that inhabits the rain forest of the Malay peninsula (Howell 1984).[13] They combine hunting and gathering with shifting agriculture, and they display the same kind of egalitarianism as pure hunter-gatherer societies. The distribution of food and other scarce items is governed by a system of superstitions known as *punen*, which roughly means "a calamity or misfortune, owing to not having satisfied an urgent desire":

> In the Chewong world, desires are most likely to occur in connection with food. If someone is not immediately invited to partake of a meal which he observes, or if someone is not given her share of any foodstuff seen to be brought back from the jungle, that person is placed in a state of punen because it is assumed one would always wish to be given a share and hence [that not being given a share would lead one to] experience an unfulfilled desire. . . . To "eat alone" is the ultimate bad behaviour in Chewong eyes, and there are several myths that testify to this. The sanction on sharing out food originates in the myth about Yinlugen Bud, who was the chief instrument in bringing the Chewong out of their pre-social state by telling them that to eat alone was not proper human behaviour. (Howell 1984, 184)

This passage suggests that the superstitions, myths, and gods of Chewong culture are intimately related to a matter of supreme practical importance—food sharing. In addition, the punen system goes beyond beliefs to include social practices that virtually assure an equal distribution of food:

> The Chewong take all possible precautions against provoking punen. All food caught in the forest is brought back and publicly revealed immediately. It is then shared out equally among all the households. The women cook it and then share the food in equal proportions among all the members of their own household. As soon as a carcass is brought back, and before it has been divided up, someone of the hunter's family touches it with his finger and makes a round touching everyone present in the settlement, each time saying "punen." . . . This is another way of announcing to everyone present that the food will soon be theirs, and to refrain from desiring it yet awhile. If guests arrive while the hosts are in the middle of a meal, they are immediately asked to partake. If they refuse, saying that they have just eaten, they are touched with a finger dipped in the food, while the person touching says "punen." (185)

Although food is virtually always in short supply, other non-foodstuffs can be scarce or common depending upon the time of year or other circumstances. The punen system is sufficiently flexible to include items only when they are scarce:

Thus bamboo for baking the tapioca bread must be shared equally among all households if the gatherer had to go very far to obtain it. If the bamboo grows close to the settlement, one may collect enough for one-self only. The difference is expressed as bamboo far away (lao tyotn) or bamboo nearby (lao duah). If the nearby river dries out and water has to be carried some distance, it again has to be shared, but daily water collection from the usual source need not be shared. . . . Even if one does not want something that has been brought back, one has to be made publicly and specifically aware of the existence of the thing, by touch if not by receipt of an actual share. (185–86)

It might seem that the injunction to honor other people's desires would lead to inappropriate sharing and the opportunity to freeload. However, this is not the case:

Once a desire has been voiced, the person who can satisfy it must immediately do so. If he refrains, the person refused will suffer the consequences of punen. But unvoiced desires are just as liable to provoke the same repercussions. In fact people hardly ever make overt requests for anything, and the fear of punen may easily have prevented people from requesting gifts from me. I can recall only one instance when I was solicited to give. An old woman, Mag, asked me to give her a whetstone, which I duly did. The rest of the Chewong, when they heard what Mag had done, commented unfavourably on her behaviour. Of course, if I had refused, it would have been because of my failing to satisfy her desire that Mag would have been bitten by a tiger, snake, or whatever. To desire for oneself can be seen to be bad on two counts. First, it is an overt emotion. Secondly, it emphasizes the individual at the expense of the social. (185)

The calamity that awaits someone who has been placed in a state of punen takes the form of an attack by a tiger, a snake, or a poisonous millipede. Moreover, these animals have spirit forms that can inflict other misfortunes such as disease or physical injury. Thus, virtually any misfortune can be used as "evidence" of a previous transgression. Immunity from disproof might seem like a weakness from a narrow scientific perspective, but it can be a strength for a social system designed to regulate human behavior.

This example gives a preview of some of the major themes that will occupy us in future chapters. The Chewong engage in many activities that would qualify as altruistic if performed voluntarily and without any system of social controls. Benefits are not meted out according to a narrow calculus of genealogical relatedness or likelihood of return gain. Instead, the group is united by a system of beliefs and practices that is essentially moral in tone. There is right conduct and wrong conduct and the latter invites punishment, not only by animals and their spirits but also by disapproving Chewong.[14] The moral system makes it hard to identify what counts as altruistic, but it obviously discourages behaviors that "emphasize the individual at the expense of the social"—to use Howell's words. I do not mean to imply that the system works perfectly and eliminates all self-serving behaviors. However the system appears designed for this purpose and by Howell's account works fairly well. If we conduct a thought experiment in which the system is eliminated, it is likely that food sharing would decline. The system has an otherworldly side that superficially appears irrational and dysfunctional. What simple people, to fear that unfulfilled desires invite attacks from animals and their spirits! Yet, a closer look reveals an impressive functionality with respect to the most practical aspects of Chewong life. We would be foolish to attribute function to each and every nuance of Chewong society, but we would be equally foolish to dismiss the possibility of function altogether. Remember that we are striving for the middle ground.

INNATE PSYCHOLOGY OF MORAL SYSTEMS

It is clear that our understanding of human social evolution must be brought within the paradigm of major transitions. We must go beyond kin selection and reciprocity to model a complex regulatory system that binds members of a group into a functional unit. Evolutionary biologists have only begun this endeavor, so it is important to emphasize the tentative nature of their conclusions. In the introduction I said that this book is about science in motion, full of inconsistencies and loose ends that will only be resolved in the future. Nevertheless, it is even more interesting to describe the game in progress than to report the final score. In this spirit, we need to reconcile two seemingly contradictory facts: the fact that moral systems require innate psychological mechanisms, and the fact that they can rapidly evolve by cultural evolution.

Beginning with psychological mechanisms, there is a long tradition in the human sciences of trying to explain as much as possible with a few general principles, such as operant conditioning or rational choice, as if the laws of behavior are like the laws of physics. In contrast, evolutionary psychologists such as Cosmides and Tooby (2001; see also Tooby and Cosmides 1992) stress that the mind is not a single general-purpose organ but a collection of many organs that adapt organisms to specific aspects their environments. Migratory birds stare at the night sky as nestlings and learn the center of rotation, which they use as adults to travel north and south (Emlen 1975). The neural circuitry that makes this possible evolved in migratory birds but not in other species. It solves only one problem of survival and reproduction. Other problems such as feeding and mating require different specialized circuits. The desert ant *Cataglyphis bicolor* takes a meandering route in search of food but carries it in a straight path back to the colony. Its tiny mind actually represents the meandering path as a series of vectors with known length and direction with respect to the sun, which enables the direction of the starting point to be "calculated" when it is time to return (Wehner and Srinvasan 1981).[15] This is only one specialized circuit that is packed into the tiny mind of the ant; other circuits are required to cope with other problems of survival and reproduction. The ability to navigate by the stars or to dead reckon by the sun appear miraculous to us because they exceed our own ability, at least without extensive training. However, our minds are also packed with specialized circuits that enable us to solve our own problems of survival and reproduction as naturally as celestial navigation in birds and dead reckoning in ants. Psychologists should be trying to identify and understand these specialized circuits rather than pretending that human behavior can be derived from a few law-like mechanistic principles.

Even without knowing the details, this basic conclusion almost certainly applies to our ability as a species to form into functional groups unified by moral systems. Like adaptations in other species, it requires a specialized, genetically evolved cognitive architecture of its own. A number of authors have speculated on the design features required for a moral system to work, including conformity (Boyd and Richerson 1985; E. O. Wilson 1998), docility (Simon 1990), detection of cheating (Cosmides and Tooby 1992), punishment of cheating (Boyd and Richerson 1992), symbolic thought (Deacon 1998), explicit consensus decision making (Boehm 1996), and so on. Much more work is required to refine these possibilities, but the general expectation is that small groups of people are psychologically prepared to bind themselves into functional units. Take one hundred people from anywhere, place them on a deserted island where they have a reason to work together, and they will make a pretty good job of it. Perhaps they will split into two groups of fifty that try to exterminate each other, but whether the common problem is a hostile environment or a hostile group, working together as a group comes naturally in our species, just as celestial navigation comes naturally to migratory birds and dead reckoning to desert ants.

It is important to stress that the phrase "comes naturally" does not imply that the underlying mechanisms are simple. Vision comes naturally but requires an amazing array of innate cognitive mechanisms. These mechanisms interact during development with features of the environment that are so reliable that all normal people can see without having to think about it. The cognition is automated and takes place beneath our conscious awareness. Similarly, our ability to function as groups may require sophisticated cognitive mechanisms that appear effortless only because they are automated. Decades were required to understand the neurobiology of vision, and a similar effort may be required to understand the neurobiology of moral systems.

The concept of an innate psychology of functional groups is not just a radical conjecture of evolutionary biologists. It is supported by some of the most distinguished research programs in the social sciences. One of the most famous studies in social psychology is the robber's cave experiment (Sherif et al. 1961), in which boys at a summer camp spontaneously formed into warring tribes which nevertheless could be brought together into a single cooperative group by confronting them with a common problem. A branch of social psychology known as social identity theory shows how easily people think of themselves as members of groups, especially in opposition to other groups (Abrams and Hogg 1990, 1999). Social dilemma experiments demonstrate the fragility of cooperation in the absence of punishment but the ease with which it is achieved when the opportunity for punishment is allowed (e.g., Ostrom et al. 1994). A book aptly entitled *Order without Law* (Ellickson 1991) shows how people spontaneously establish, enforce, and largely abide by social norms in the absence of a formal legal system. Tocqueville, the French social theorist who observed American democracy with such insight, was equally perceptive about small-scale human society in general when he said that "the village or township is the only association which is so perfectly natural that, wherever a number of men are collected, it seems to constitute itself" ([1835] 1990, 60). There is great opportunity for a synthesis on this subject between the established branches of the social sciences and evolutionary biology, upon which all functional explanations must ultimately rest.

Applying these insights to the study of religion, we should not think of religion as a purely cultural invention or as something that can be derived from a few law-like principles. Organisms of all sorts require a complex and specialized physiology to coordinate their parts in just the right way to survive and reproduce in their environments. We should think of the psychological mechanisms activated by religion as physiological in this sense.

CULTURAL EVOLUTION OF MORAL SYSTEMS

The second basic fact that we must understand from an evolutionary perspective is that moral systems include an open-ended cultural dimension in addition to an innate psychological dimension. Our genetically evolved minds make it possible to have a moral system, but the specific contents of moral systems can change within groups and vary widely among groups, with important consequences for survival and reproduction. Far from leading to the caricature of genetic determinism that limits the capacity for change, our innate psychology creates a capacity for change by setting in motion a process of cultural evolution.

Although culture has for many decades been envisioned as an evolutionary process, there is little agreement about its precise nature, importance, or relationship to genetic evolution. The most severe critics of sociobiology rely upon culture as an alternative, which they think can be studied without reference to biology (e.g., Sahlins 1976). Some biologists regard culture as a handmaiden of genetic evolution that evolves the same phenotypic adaptations, only faster (e.g., Alexander 1979, 1987). Other biologists try to decompose culture into gene-like units that do not necessarily benefit their human hosts (e.g., Dawkins 1976; Blackmore 1999). Instead, they can act more like disease organisms as they spread from head to head.[16]

To find our way through this wilderness of possibilities, let's begin with the image of the mind as a collection of specialized organs, which I emphasized in the previous section. The reason the organs must be specialized is supposedly because general-purpose cognitive organs are not possible (Tooby and Cosmides 1992; Cosmides and Tooby 2001). The first artificial intelligence researchers naively thought that they could build smart general-purpose learning machines, but they soon discovered that the only way to make a machine smart is to make it specialized for a particular task. Chess playing computers are smart at playing chess but can't do anything else. Similarly, the neural circuit that enables migratory birds to learn the axis of rotation of the night sky can't do anything else. In general, the world is so full of stimuli and possible responses to stimuli that a smart machine, or a smart cognitive organ, must be very selective about its perception and use of information. As a Chinese proverb states, a wise man knows what to ignore.

These are valid observations, but they have been used by Cosmides, Tooby, and others most closely associated with the term "evolutionary psychology" to build an incomplete picture of the human mind as like a juke-box, in which the records are preevolved cognitive modules and the environment is the button-pusher.[17] Whatever module that is playing at the moment requires information about the environment in the same way that a computer program requires the input of information. For example, suppose that you are riding on a bus and a member of the opposite sex sits across from you. Out pops your mate choice module, which causes you to scan the person for information relevant to their quality as a mate; their age, health, resources, availability, and so on. A degree of learning takes place during this process but it is highly formulaic. Just as a tax preparation program tells you what you owe after "learning" your income, business expenses, and so on, the mate choice module tells you how attractive the person is after "learning" his or her salient qualities.

Once again, the problem with this portrayal of human mentality is not that it is wrong but that it is partial (Wilson 1994, 1999b, 2002). Against the background of all-purpose learning theory (what Cosmides and Tooby call the "Standard Social Sciences Model" or SSSM), it is full of insights and possibilities. By itself, however, it seems to deny learning, development, and cultural change as open-ended processes. Learning, as we have seen, becomes highly circumscribed information processing. Development becomes the switching on and off of modules during various stages of the life cycle. Culture becomes a reflection of individual behavioral flexibility. If people in different locations experience different environments, different modules will be triggered which in turn will lead to different behaviors. We might call these differences cultural but they have nothing to do with socially transmitted information per se.[18]

Self-described evolutionary psychologists might complain that I am caricaturing their position. Perhaps I am—but not by much. A glance at the index of David Buss's *Evolutionary Psychology: Toward a New Psychology of the Mind* reveals that only one page is devoted to "learning," seven pages to "development," and seven pages to "culture." The contents of these pages are as I have described in the previous paragraph (see Wilson 1999b for a detailed critique). The bulk of the book and the field that it represents employs the same algorithm again and again: For any particular feature of human behavior and psychology, try to understand it as a genetically evolved adaptation to a feature of the ancestral environment. Then try to imagine the psychological mechanism as a specialized module.

In this algorithm, everything that has taken place since the advent of agriculture counts for nothing, other than as a source of maladaptive behavior. Why do we overeat? Because we are genetically adapted to crave fats and sugars in the food-poor ancestral environment, which places us at the mercy of every fast-food restaurant in today's food-rich environment (Stevens and Price 1996). Why are we displeased with our mates? Because the media bombards us with images of the most beautiful people on earth, making even above-average people appear like toads by comparison. In the ancestral environment, the comparison pool would be at most a few hundred individuals (Buss 1999). Why do we often behave stupidly in psychological experiments? Because we are not given information in the form of frequencies, which is how we were designed to receive information in the ancestral environment (Gigerenzer and Hoffrage 1995). The algorithm provides no explanation for why the modern environment became so different from the ancestral environment, but merely accepts the fact and tries to deal with it, like a biologist trying to study a species of rain forest lizard that has mysteriously been transported into the desert.

I must stress once again that I like these ideas, which have led to many valid insights that are new against the background of traditional psychological research. I myself have already stressed the importance of innate psychology in the study of human moral systems. My complaint is not that the algorithm is wrong but that it is partial, seeming to exclude the possibility of learning, development, culture, and other aspects of human mentality as open-ended processes. To broaden the horizon, we need to return to the question of whether cognitive processes must be specialized to be smart.

Consider the mammalian immune system. Just like the mind, it can be regarded as a collection of specialized genetically evolved mechanisms for helping us survive and reproduce in our ancestral environment. The number and sophistication of the mechanisms that comprise the immune system are mind-boggling when understood in detail. Nevertheless, the centerpiece of the immune system is an open-ended process of blind variation and selective retention. Antibodies are produced at random and those that successfully fight invading disease organisms are selected. Diseases are so numerous and evolve so fast with their short generation times that the only way to fight them is with another evolutionary process.

This comparison, between the mind and the immune system, is simple but profound in its implications.[19] It shows that genetic evolution does not invariably lead to the kind of modularity that excludes open-ended processes. Instead, it can create processes that are themselves evolutionary and therefore capable of providing new solutions to new problems. Plotkin (1994) has aptly termed these processes "Darwin machines," two words that reflect the essential components of an evolved system that includes evolution within its own structure. "Machine" indicates that the internal evolutionary process must be highly managed to lead to biologically adaptive outcomes. Antibodies that match antigens reproduce more, not by chance, but because the immune system has been constructed that way. "Darwin" indicates that the internal process remains evolutionary despite being managed, with all the implications associated with genetic evolution played out on a new stage. Adaptation to recent (not ancient) environments is perhaps the most important implication, but we should also expect the same kinds of historical contingencies and other constraining factors that cause adaptations to fall short of perfection.

Thinking of the mind as like the immune system allows us to appreciate its genetically evolved and highly specialized features without denying its open-ended potential. Against this background, we can return to the subjects of cultural evolution, moral systems, and religion. As many authors have noted, there was no single human ancestral environment but rather many environments that varied over time and space. The physical environment was exceptionally variable during our emergence as a species, due to unstable climatic conditions (Richerson and Boyd 2000). In addition, human social interactions have the same what-I-do-depends-on-what-you-do quality that marks the interactions between hosts and their disease organisms. When physical and social environments become sufficiently variable, juke-box solutions are inadequate and the only recourse is to evolve Darwin machines.

Cultural evolution can be seen in part as a Darwin machine in action, highly managed but nevertheless genuinely open-ended in its outcome. Confront a human group with a novel problem, even one that never existed in the so-called ancestral environment, and its members may well come up with a workable solution. The solution might be based on trial and error or on rational thought. However, rational thought is itself a Darwin machine, rapidly generating and selecting symbolic representations inside the head. Confront many human groups with the same novel problem and they will come up with different solutions, some much better than others. If the groups are isolated from each other, they may never converge on the best solution; evolution is not such a deterministic process. If the groups are in contact, they might compare solutions and the worst might quickly imitate the best. If convergence by imitation does not occur, then the worst might simply succumb to the best in between-group interactions. Either way, the final outcome is a degree of adaptation to the problem, without any genetic evolution taking place at all. Evolution took place, but not at the genetic level.

This description of human cultural evolution sounds so familiar that the reader might wonder if I have said anything new. Scientific progress does not always involve replacing the familiar with the counterintuitive. Some things are familiar because they are true. A good theory acknowledges the valid aspects of the familiar and goes on to achieve a new level of understanding. In this spirit, multilevel selection theory contributes at least four major insights to the "familiar" process of cultural evolution.

First, cultural evolution requires specialized mechanisms—the machine part of the Darwin machine. Familiar-sounding terms such as "trial and error," "rational thought," and "imitation" probably don't even begin to describe the number and sophistication of the mechanisms that actually guide the process of cultural evolution. More generally, thinking of cultural evolution as itself a product of genetic evolution with many sophisticated design features is anything but familiar.

Second, many of the mechanisms guiding cultural evolution take place beneath conscious awareness. We have a tendency to attribute too much importance to conscious rational thought. We imagine ourselves solving problems by explicitly thinking, talking, experimenting, imitating, and so on. These conscious processes are important agents of cultural change (Boehm 1996), but they are the tip of an iceberg of automated cognitive processes that take place beneath our conscious awareness, some of which are very sophisticated. This means that cultures can evolve to be smart in ways that are invisible to their own members.

Third, the mechanisms guiding cultural evolution can be distributed processes involving many individuals rather than being processes contained within single individuals. There is a pervasive tendency both in biology and in the human sciences to regard individuals as self-contained cognitive units. An individual might rely on outside information and might decide to cooperate with others, but it is still the individual's decision. The evolutionary justification for this claim is as follows: A brain is a group of neurons whose members interact intricately for the common good. There is a big difference between a brain and a single neuron; most of what we attribute to the mind involves the circuitry connecting the neurons, not the on/off states of the neurons themselves. The reason that neurons in a brain act for their common good is because they exist within a single individual organism that survives and reproduces as a unit. Brains are self-contained cognitive units because individuals are units of selection.

This reasoning, along with so much else that until recently appeared on solid ground, must be questioned in the light of multilevel selection theory. If the individual is no longer a privileged unit of selection, it is no longer a privileged unit of cognition. We are free to imagine individuals in a social group connected in a circuitry that gives the group the status of the brain and the individual the status of the neuron.

The concept of a group brain might seem like science fiction, but only against the background of the last fifty years of intellectual thought. As I mentioned earlier, the founding fathers of the human social sciences were fully comfortable with the idea of group organisms, complete with group minds (Wegner 1986). Moreover, modern social insect biologists have established the reality of group minds in impressive detail (Seeley 1995; Detrain et al. 1999). An example from honeybees will help to restore our intuition for our own species.

The challenges of survival for a honeybee colony are awesome when sufficiently appreciated. To function adaptively, the colony must make decisions on an hourly basis about which flower patches to visit and which to ignore over an area of several square miles, whether to gather nectar, pollen, or water, the allocation of workers to foraging vs. hive maintenance, and so on. In an elegant series of experiments, T. D. Seeley and his colleagues worked out in detail how some of these decisions are actually made (reviewed by Seeley 1995). In one experiment, a colony in which every bee was individually marked was taken deep into the Adirondack woods where virtually no natural resources were available. The colony was then provided with artificial nectar sources whose quality could be experimentally manipulated. When the quality of a food patch was lowered below alternative patches, the colony responded by shifting workers away from the patch, yet individual bees visited only one patch and therefore had no frame of comparison. Instead, individuals contributed one link to a chain of events that allowed the comparison to be made at the colony level. Bees returning from the low-quality patch danced less and themselves were less likely to revisit. With fewer bees returning from the poor resource, bees from better patches were able to unload their nectar faster, which they used as a cue to dance more. Newly recruited bees were therefore directed to the best patches. Adaptive foraging decisions were made by a decentralized process in which individuals acted more as neurons than as decision-making agents in their own right. Even the physical architecture of the hive, such as the location and dimensions of the dance floor, honeycomb, and brood chambers, has been shown to play an important role in the cognitive architecture of adaptive decision making at the group level.

A more recently documented example of group cognition in honeybees involves the selection of a new site (Seeley and Buhrman 1999). When the colony becomes large enough to split, the old queen departs with about half the workers and forms an exposed mass of bees called a swarm from which scouts emanate to search for suitable sites. Scouts returning from different sites interact on the surface of the swarm in such a way that the best site is chosen. As with foraging for nectar, however, no single scout visits more than one site, so decision making is a distributed process that requires a group of bees interacting in just the right way. This example is instructive because it is equivalent to a "best-of-n" decision-making algorithm that is also employed by individual humans (Payne et al. 1993). The algorithm is the same, regardless of whether it involves a group of neurons or a group of bees.

Social insect colonies have taken the concept of group minds and group organisms beyond the realm of mysticism and science fiction. In addition, it is entirely likely that the same concepts apply to our own species (Wilson 1997). As much as we might laud the individual human mind, its capacity is vastly exceeded by the demands of language and culture that are the hallmark of our species and that evolved in tandem with human brain evolution. Theories of human evolution frequently emphasize various forms of cooperation in the context of physical activities (e.g., hunting, intergroup warfare); why not also in the context of mental activities? If widespread cognitive cooperation did evolve in our species, we need be no more aware of the role that we play in the group mind than honey bees as they perform their waggle-dance.

The fourth insight that multilevel selection theory contributes to the "familiar" process of cultural evolution is that such evolution takes place largely at the group level. Cultural evolution is not merely a handmaiden of genetic evolution but changes the parameters of the evolutionary process, favoring traits that would not evolve by genetic evolution alone (Boyd and Richerson 1985; Boehm 1999; Wilson and Kniffin 1999). To understand the significance of this statement, consider a genetic mutation that occurs in a large population that is subdivided into groups. There is now genetic variation both within groups (the single mutant vs. everyone else in the group) and between groups (a group with a single mutant vs.

many groups with no mutant). The gene can be favored by group selection if the mutant individual single-handedly increases the fitness of its group, relative to other groups, but this advantage must be weighed against the fitness of the individual, relative to the other members of the same group. The group effect is likely to be more powerful if the groups are small (e.g., 1 out of 5) than large (e.g., 1 out of 1,000). Group selection would work better if we could concentrate mutants into a single group, but it is not obvious how this can happen if the mutant gene is at a low frequency in the total population. These are some of the limiting factors that set the tone of the group selection debate in the 1960s.

Now consider a cultural mutation; a new belief or practice that arises in one group by chance, rational thought, or any other process. Unlike the genetic mutation, the cultural mutation need not remain at a low frequency within the group. A variety of factors can swiftly make the mutant behavior the majority or even the exclusive behavior practiced by the group. Perhaps members think explicitly about the new belief or practice, appreciate its wisdom, and establish it as the new norm. Perhaps they follow it "mindlessly" because it is espoused by a charismatic member of the group. If the mutant behavior is altruistic (when practiced voluntarily), its costs and benefits can be modified by the full panoply of social control mechanisms available in our species. These factors can cause a rare behavior to become common even in very large groups. In short, cultural evolution increases the potency of selection among groups and decreases the potency of selection within groups, compared to what would be expected on the basis of genetic evolution alone. This is not an inevitable consequence of cultural evolution, but it is how cultural evolution appears to work in our species—a design feature of a Darwin machine.[20]

Applying these insights to the study of religion, we should think of religious groups as rapidly evolving entities adapting to their current environments. Religions appeal to many people in part because they promise transformative change—a path to salvation. The word evolution means change, so it would seem that evolution and religion share much in common. It is unfortunate that evolution is so often associated with genetic evolution, a slow process that gives the impression of an incapacity for change over the time scales that matter most to living people struggling with their problems. When we expand our view of evolution to include all Darwinian processes, we can begin to see how religions actually can produce transformative change, even from a purely evolutionary perspective.

MODERN HUMAN GROUPS AS ADAPTIVE UNITS

Most modern societies are vastly different from the hunter-gatherer groups of the ancient past. Although significant genetic evolution can occur in a small number of generations (Endler 1986; Weiner 1994), the basic genetic architecture of the human mind has probably not changed much since the advent of agriculture approximately ten thousand years ago. As we have seen, when evolution is interpreted too narrowly as genetic evolution, all of recorded history becomes a mystery from an evolutionary perspective, something that happened but cannot be explained. The best we can do is try to understand how the stone-age mind is likely to react to the strange new world for which it is not prepared. An expanded view of evolution allows us to interpret recorded history as a fossil record of cultural evolution in action. As with genetic multilevel selection, a cultural variant can spread at the expense of other variants within a group, or by causing its group to spread at the expense of other groups.

Superficially, large-scale human societies appear much less egalitarian than hunter-gatherer groups, but the apparent inequities can be interpreted in two very different ways. On the one hand, social control mechanisms are probably strongest in small groups in which everyone knows and depends on everyone else. Many inequities that exist in large-scale societies are therefore exactly what they seem—some individuals profiting at the expense of others within the society. These should not be interpreted as group-level adaptations but rather as individual-level adaptations with consequences that are often dysfunctional at the society level. On the other hand, purely from the group-level functional standpoint, societies must become differentiated as they increase in size. Thirty people can sit around the campfire and arrive at a consensual decision; thirty million people cannot. It is therefore an open question whether extreme status differences and other seeming inequalities in large-scale societies represent domination pure and simple or rather design features that enable the society to function at a large scale, especially in competition with other societies. There can be little doubt that size itself can be a group-level adaptation. Larger societies tend to replace smaller societies unless their larger size is offset by problems of coordination and internal conflicts of interest. It is possible to imagine major transitions occurring in cultural evolution, in which smaller groups coalesce into larger groups, just as for long-term biological evolution.

Our progress so far can be summarized as follows. Organismic groups do not automatically evolve but require a process of group selection.

Group selection can be a potent evolutionary force, despite its widespread rejection during the age of individualism. In fact, the organisms of today are the social groups of past ages, which have become so functionally integrated that we see the whole more than the parts. These developments in evolutionary biology make the organismic view of human society a legitimate possibility—at least to a degree. Human societal organisms rely critically on moral systems to define appropriate behaviors and to prevent subversion from within. Moral systems have an innate psychological dimension but also an open-ended dimension that allows human history to be seen as a fast-paced evolutionary process with cultural rather than genetic mechanisms of inheritance.

Two more issues need to be discussed to complete our survey of evolutionary concepts relevant to the study of religion. First, we must take a closer look at the concept of fitness. Second, we must ask why morality is so often expressed in the form of religion, which seems so different from other modes of thought.

WHAT CONSTITUTES FITNESS?

Studying religion (or any other subject) from an evolutionary perspective requires a clear definition of fitness. Sometimes it is obvious how an organism must be structured to survive and reproduce in its environment. Flight requires certain aerodynamic shapes, and the efficiency of a bird's wing can be measured with the same precision as the efficiency of an airplane's wing. The efficiency of behaviors can also be measured with precision. Traveling salesmen must move between cities to sell their wares. There are many ways to construct a path through a number of cities, in fact so many ways that even large computers cannot evaluate them all. Nevertheless, we can easily test whether traveling salesmen choose relatively efficient paths. We can also test whether hummingbirds choose relatively efficient paths on their trips from flower to flower.

It is important to remember that the evidence for design in nature is so compelling that it cries out for an explanation. The great question in Darwin's day was not "Is there any function in nature?" but "What explains all the function that we see in nature?" Thus, at a certain basic level the question "what constitutes fitness" can have a satisfying and highly intuitive answer. However, a closer look reveals a host of complicating factors that can make fitness difficult to define and study (Michod 1999a).

One of the complicating factors has already been discussed and forms the heart of this book: Fitness is a relative concept. It doesn't matter how well an organism survives and reproduces. It only matters that it survives and reproduces better than alternative types of organisms. Males of some species are adapted to kill infants, which enables them to mate with the infants' mothers faster than they could otherwise (Van Schaik and Janson 2000). This behavior is not adaptive for the infants, the mothers, the group, the species, or the ecosystem. It is adaptive only for the males, compared to males who behave otherwise. Nevertheless, these males must be regarded as fit from an evolutionary perspective. Some species of bees have evolved to drink nectar without becoming dusted with pollen, by chewing a hole in the base of the flower. This behavior is not adaptive for the flower-bearing plants or even the bee species, which depends upon the plants for its long-term survival. It is adaptive only for the individual bee, compared to bees who behave otherwise. It is hard to avoid a feeling of moral revulsion at calling such behaviors fit when they are so destructive to other organisms and even the "fit" organism itself over the long term. As we have seen, group selection is a partial solution to this problem. Groups of males who do not kill each other's infants might survive and reproduce better than other groups. The feeling of moral revulsion that I just described can itself be explained as part of the innate psychology of moral systems that evolved by group selection to suppress self-serving behaviors in our own species. But alas, group selection merely takes us out of the frying pan of within-group interactions and into the fire of between-group interactions. Those groups of males who do not kill each other's offspring might well kill the offspring and appropriate the females from other groups (Wrangham and Peterson 1997).

These points must be kept firmly in mind when we proceed to our study of religion. Whenever I strike up a conversation about religion, I am likely to receive a litany of evils perpetrated in God's name. In most cases, these are horrors committed by religious groups against other groups. How can I call religion adaptive in the face of such evidence? The answer is "easily," as long as we understand fitness in relative terms. It is important to stress that a behavior can be explained from an evolutionary perspective without being morally condoned. Immoral behaviors almost invariably benefit the immoral individual or group; why else would immorality be a temptation? Evolution is not required to tell us something so basic. Religious discussions of self-will are a breath away from evolutionary discussions of self-interest. Open-minded religious believers are perfectly aware that solving the problem of self-will within religious groups can lead to even greater problems of group-will with respect to other groups. These parallels between religious and evolutionary thought are not coincidental; they both spring from the fundamental problem of social life and its partial solution that lies at the heart of religion and which can be explained by multilevel selection theory.

Not only is fitness a relative concept, but it is also a local concept. The English system of measuring in feet and inches is inferior to the metric system, but it persists in certain populations because it is common. The cost of switching to the metric system outweigh the benefits, at least over the short term. This is known as a majority effect, and examples abound in both biological and cultural evolution. IBM-compatibles have an advantage over Apple computers and Microsoft Word has an advantage over other word processing programs because of the majority effect. These examples do not violate the principle of evolution as a fitness-maximizing process but simply illustrate its local nature, which is often illustrated with the metaphor of an adaptive landscape.[21] Imagine the English measurement system as a meager hill of low fitness and the metric system as a taller hill of high fitness. Evolution is a hill-climbing process, but if it starts out on the slope of the meager hill, all it can do is climb to the top of that hill. Moving from a short hill to a tall hill requires crossing a valley of low fitness and is actually resisted by the evolutionary process. The more rugged the adaptive landscape, the more an evolving system will reflect its original starting point (the particular hill upon whose slope it landed) and will fail to find the best global solution.[22]

A third complication involves mechanisms of inheritance. The sickle cell gene (S) in humans is an adaptation to malaria, but it does not spread to fixation because it is only advantageous in heterozygotic form (AS). As a homozygote (SS) it leads to debilitating anemia. The result is a genetic polymorphism in which some individuals are protected from malaria (AS) while many others are either unprotected (AA) or suffer from a genetic defect (SS). Some adaptation! In general, the more we complicate the mechanisms of genetic inheritance, making genes fit in some combinations but unfit in others, the messier the process of adaptation becomes.

A fourth complication involves modes of transmission. In diploid organisms such as ourselves, most genes exist in the autosomes and are inherited from both parents, but some genes exist in the cytoplasm and are inherited only from the mother. All cytoplasmic genes in males are doomed to extinction because they will not enter the male's sperm. A mutant gene that causes females to raise daughters instead of sons will be favored by natural selection if the gene is cytoplasmic but not if it is autosomal. Similarly, a gene that causes males to raise sons instead of daughters will be favored if it is located on the y-chromosome, which is transmitted only through males. These conflicts of interest among genes reveal that even individual organisms fall short of the ideal of internal harmony implied by the word "organism" (Pomiankowski 1999). The fact that fitness depends on the mode of transmission has important implications for models of cultural evolution, which include many possible modes of transmission (Boyd and Richerson 1985).

A fifth complication involves distinguishing the product of natural selection from the process. An adaptation is a product of natural selection and is expected to be well designed with respect to the environment. However, the process of natural selection involves many failures for each success. Like laws and sausages, the manufacture of adaptations is not a pretty sight! Religious experiments that fail are not an argument against evolution, if we are observing the process in addition to the product. The question is whether religious experiments that succeed do so on the basis of their properties, and whether these properties are transmitted (with modification) to subsequent religions.

All of these complications (and others) are important and must be kept in mind as we proceed to our study of religion. However, they should not obscure the progress that can be made at a basic level. Religions are often concerned with the necessities of life—food, shelter, health, safety, marriage, child development, social relations of all sorts. These are so obviously related to survival and reproduction that, at least to a first approximation, we needn't puzzle over the details any more than Darwin needed to puzzle over the details of a thick beak useful for cracking hard seeds. In addition, it is often clear enough whether people are obtaining the necessities of life at the expense of others or by coordinating with others. The fundamental problem of social life and the role of religion in its (partial) solution are too basic to be obscured by the many complications surrounding the concept of fitness, or so I will try to show.

THE RELIGIOUS EXPRESSION OF MORALITY

So far I have said much about evolution, human evolution, morality, and culture, but little about religion per se. Even if we accept that moral systems enable human groups to function as adaptive units, what accounts for the religious expression of morality? Why can't people just talk about right and wrong in practical terms without appealing to supernatural agents and other beliefs that to a nonbeliever seem detached from reality?

Religion attracts the attention of scientists (and often the scorn of nonbe-lievers) in part because it seems to flaunt the canons of scientific thought. For many people, the otherworldly nature of religion is more interesting and important to explain than its communal nature.

One possibility is that religions are naive scientific theories, attempts by simple people to understand their complex world that just happen to be false. Religious folk should abandon their beliefs in the face of superior knowledge and if they don't they are being irrational. This idea surfaces again and again in casual discussions and also forms the basis of more formal theories of religion (Frazer 1890; Tylor 1871). However, it fails to fit the facts. In the first place, people in all cultures—even the most "primitive"—possess the foundation of scientific thought: a sophisticated factual understanding of their world and the ability to reason on the basis of evidence (Malinowski 1948; Boehm 1978). They do not live in an oth-erworldly fog in all respects. The fog—if that is what it deserves to be called—only descends in some contexts. In the second place, there is no evidence that scientific understanding replaces religious belief in modern cultures. America has become more religious over the course of its history, not less, despite the influence of science and engineering (Finke and Stark 1992). A very high proportion of scientists themselves profess a belief in God and participate in organized religions (Stark and Finke 2000). Clearly, we must think of religious thought as something that coexists with scientific thought, not as an inferior version of it.

A proper understanding of epistemology from an evolutionary per-spective can shed light on this issue. Before Darwin, the human ability to know (i.e., to accurately perceive the properties of the external world) could be explained as a gift from God. After Darwin, numerous philoso-phers and biologists tried to place epistemology on an evolutionary foun-dation by saying that the ability to know is adaptive (Bradie 1986). Those who did it well survived and reproduced while those who did it poorly were not among our ancestors. This argument has an element of truth; clearly, I need to accurately perceive the location of a rabbit to hit it with my throwing stick. However, there are many, many other situations in which it can be adaptive to distort reality (Wilson 1990, 1995). Even mas-sively fictitious beliefs can be adaptive, as long as they motivate behaviors that are adaptive in the real world. At best, our vaunted ability to know is just one tool in a mental toolkit that is frequently passed over in favor of other tools—just as we observe in all cultures, including our own.

From this perspective, we should expect moral systems to frequently depart from narrow reasoning on the basis of factual evidence. Once this kind of reasoning is removed from its pedestal as the only adaptive way to think, a host of alternatives become available. Emotions are evolved mechanisms for motivating adaptive behavior that are far more ancient than the cognitive processes typically associated with scientific thought. We might therefore expect moral systems to be designed to trigger powerful emotional impulses, linking joy with right, fear with wrong, anger with transgressions. We might expect stories, music, and rituals to be at least as important as logical arguments in orchestrating the behavior of groups. Supernatural agents and events that never happened can provide blueprints for action that far surpass factual accounts of the natural world in clarity and motivating power. These otherworldly elements of religion cannot completely eclipse scientific modes of thought, which are superior in some contexts, but the reverse statement is equally true.

A second example of hunter-gatherer morality will make our discussion less abstract. The following passage from Turnbull (1965, 180) describes how men of the Mbuti tribe, which inhabits the rain forest of equatorial Africa, make decisions on a consensus basis:

Njobo was an undisputed great hunter, knew the territory as well as anyone and had killed four elephants single-handed. He was a good enough Mbuti not to attempt to dominate any hunting discussion in the forest, merely to take a normal part. If he ever appeared to be overly aggressive or insistent he was shouted down and ridiculed, although highly popular. He was also the one chosen to represent the band to the villagers. Ekianga, on the other hand, was less generally popular and was the source of some friction, having three wives (one the sister of another prominent member of the band), but he was a fine hunter, endowed with exceptional physical stamina, and he too knew the territory well. Even at the height of his unpopularity he was one of the most effective "leaders" of the hunt. So was Nikiabo, a youth who had achieved some notoriety by killing a buffalo when barely out of childhood. Although a bachelor, he had a net of his own and took a prominent part in all hunting discussions. Makubasi, a young married hunter, was also accorded special respect because of his hunting prowess and his physical strength, combined with his knowledge of the territory. But while these four can be singled out as exceptional, they could either separately or together be outvoted by the rest of the hunters. On such occasions they were compelled either to give their assent to the popular decision or to refrain from joining the hunt that day. None of them had the slightest authority over the others. Nor was any moral pressure brought to bear in influencing a decision through personal considerations or respect. The only such moral consideration ever mentioned was that when the band arrived at a decision, it was considered "good" and that it would "please the forest." Anyone not associating himself with the decision was, then, likely to displease the forest, and this was considered "bad." Any individual intent on strengthening his own argument might appeal to the forest on grounds that his point of view was "good" and "pleasing"; only the ultimate general decision, however, would determine the validity of his claim.

This passage provides a fine illustration of hunter-gatherer egalitarianism, in which the will of the group prevails over the will of the strongest individuals, at least some of whom would gladly become more dominant in the absence of social controls. Their decision about where to hunt almost certainly relied upon practical reasoning on the basis of detailed factual knowledge that we associate with scientific thought. Nevertheless, this mode of thought blended seamlessly with belief in a forest capable of experiencing pleasure, which happens to correspond to the welfare of the group. It is interesting to ask why Mbuti thought took this "irrational" turn when discussion of shared interest could have remained on a purely pragmatic and "rational" plane. However, the answer to this question must be based primarily on the adaptedness of the actions motivated by alternative modes of thought. Once the reasoning associated with scientific thought loses its status as the only adaptive way to think, other forms of thought associated with religion cease to be objects of scorn and incomprehension and can be studied as potential adaptations in their own right.

Surveying the View from Evolutionary Biology

The main purpose of this book is to treat the organismic concept of religious groups as a serious scientific hypothesis. The contribution of this chapter is to review the relevant concepts in evolutionary biology. We have covered an enormous amount of ground and touched upon many complex issues, but the basic argument can be briefly and simply stated: Natural selection is a multilevel process that operates among groups in addition to among individuals within groups. Any unit becomes endowed with the properties inherent in the word organism to the degree that it is a unit of selection. The history of life on earth has been marked by many transitions from groups of organisms to groups as organisms. Organismic groups achieve their unity with mechanisms that suppress selection within groups without themselves being overtly altruistic. Human evolution falls within the paradigm of multilevel selection and the major transitions of life. Moral systems provide many of the mechanisms that enable human groups to function as adaptive units. Moral systems include both an innate psychological component and an open-ended cultural component that enables groups to adapt to their recent environments. Belief in supernatural agents and other elements that are associated specifically with religion can play an important role in the structure and function of moral communities.

In the Introduction I stated that science works best when it tests among well-framed hypotheses that make different predictions about measurable aspects of the world. Evolutionary theory offers not one but several hypotheses about religion that differ in their predictions, as shown in table 1.1. The most important division concerns adaptive vs. nonadaptive explanations. Among adaptive explanations, religion might be a group-level adaptation (the thesis of this book; see also E. O. Wilson 1998), an individual-level adaptation (Alexander 1987), or a cultural parasite that spreads at the expense of both human individuals and groups (Dawkins 1976; Boyer 1994, 2001; Blackmore 1999). Among nonadaptive explanations, religion can be interpreted as an adaptation to past environments that has become maladaptive in modern environments or as a byproduct of evolution whose function, if any, is secondary to a more general adaptive design (Guthrie 1995; Boyer 2001). In a famous essay that criticized evolutionary biologists for relying too heavily on the concept of adaptation, Gould and Lewontin (1979), themselves eminent evolutionary biologists, said that many biological structures are like a spandrel, which is the area created by two adjoining arches. Arches are clearly functional in the design of a building but spandrels are merely the byproducts of arches. These spaces are sometimes used for artistic purposes, but their "function" is secondary at best. Similarly, noses hold up our glasses, but this "function" of noses is clearly secondary to breathing and smelling. One example of a spandrel hypothesis for religion is that self-awareness evolved by natural selection for its survival value, with the unfortunate byproduct that self-aware individuals can foresee their own deaths. Religion might then have arisen to help allay the fear of death, a secondary adaptation that can be understood only in the context of a more primary adaptation (self-awareness).

The hypotheses in table 1.1 are used by evolutionary biologists to organize the study of many subjects. When applied to religion, they make vastly different predictions that can be pitted against each other with em-

---

Table 1.1 Evolutionary Theories of Religion

---

1. Religion as an Adaptation
   1.1 Religion as a group-level adaptation
   1.2 Religion as an individual-level adaptation
   1.3 Religion as a cultural "parasite" that often evolves at the expense of human individuals and groups
2. Religion as Nonadaptive
   2.1 Religion as an adaptation to past environments, such as ancestral kin groups, that is maladaptive in modern environments, such as large groups of unrelated individuals
   2.2 Religion as a byproduct (or "spandrel") of genetic or cultural evolution

---

pirical tests. A religion designed to allay the fear of death will be different than a religion designed to promote the common good, which in turn will be different than a religion designed as a tool of within-group exploitation, which in turn will be different than a religion for which the word "design" makes no sense at all. There is ample empirical information about religion to discriminate among such different hypotheses, resulting in genuine scientific progress. In short, even if my hypothesis turns out to be incorrect, religion can and should be a subject of mainstream evolutionary research.[23]

Throughout this chapter I have stressed the need to achieve a middle ground in which groups can and do evolve into adaptive units, but only if special conditions are met. It is definitely not my intention to adopt an extreme stance on religious groups as organismic units. I think that group selection can explain much about religion, but by no means all. Evolution is a notoriously messy process that defies single explanations. Nothing is perfectly adaptive or a product of only one level of selection. Even individual organisms have not entirely solved the fundamental problem of social life within themselves. All of the hypotheses listed in table 1.1 have at least some merit; our challenge is to discover their relative importance. Also, religion is not a single trait; it is a heterogeneous set of traits that might require different explanations. Finally, science involves comparing hypotheses, so all must be considered even if only one prevails.

Thus, in many respects I will be evaluating all of the hypotheses listed in table 1.1. However, the group selection hypothesis deserves special recognition, in part because it explains so much about religion and in part because it has been so maligned and neglected. I believe that future generations will be amazed at the degree to which groups were made to disappear as adaptive units of life in the minds of intellectuals during the second half of the twentieth century. Against this background, thinking of groups as organisms whose function requires a complex and highly organized physiology is so new, and leads to so many predictions that would not be forthcoming otherwise, that it deserves recognition even if it turns out to be only partially true.

There is another reason why group selection explains the essence of religion in a way that the other hypotheses do not. When religious believers describe their church as like a body or a beehive, they are speaking idealistically. According to the Hutterite passage quoted at the beginning of the introduction, it is *true* love that means growth for the whole organism. Real churches only approach the ideal of true love, but their failures are interpreted as corruptions and aberrations of religion, not as a part of religion itself. When we study religion as it is actually practiced, we see group selection contending with, and not always prevailing against, other strong forces. When we study religion as it is idealized, we see something much closer to an expression of what would evolve by pure group-level selection.

Our study of religion will repeatedly draw upon the evolutionary themes that were developed in this chapter. First, however, we must survey the view from the social sciences, where religion has already been studied for well over a century.

CHAPTER 2

THE VIEW FROM THE SOCIAL SCIENCES

A religion is a unified system of beliefs and practices relative to
sacred things . . . which unite into one single moral community
called a Church, all those who adhere to them.
—Durkheim [1912] 1995, 44

What do human beings get from religion? According to Rodney
Stark and William Bainbridge, they get what they cannot have.
—Buckser 1995, 1

The scientific study of religion has traditionally been the province of anthropologists, sociologists, and psychologists. Interest in religion began
with the framers of these disciplines (e.g., Tylor 1871; Frazer 1890; Durkheim 1912; James 1902) and continues with today's most sophisticated
practitioners. Theories of religion in the social sciences have relied upon
evolution to varying degrees, although the historical trend has been in
the direction of less, not more. An impressive body of information has
accumulated that can be used to test hypotheses, including evolutionary
hypotheses that were not in mind when the data was being gathered.

As a newcomer, any modern evolutionary approach to religion must
prove itself against these older traditions. One extreme possibility is that
evolutionary theory triumphs over the social sciences, which totally
missed the boat. Another extreme possibility is that evolutionary theory
merely rediscovers what social scientists have long known. To continue
the nautical metaphor, an evolutionary theory of religion might be like a
rowboat joining a fleet of battleships, with nothing new to offer. A third

extreme possibility is that evolutionary theory fails to explain the nature of religion; the battleships sink the rowboat.

The real situation is more complex and interesting than any of these extremes. The hypothesis that I developed in chapter 1 is close to a position in the social sciences known as functionalism, which was far more popular during the first half of the twentieth century than today. In fact, the demise of functionalism in the social sciences bears an eerie resemblance to the demise of multilevel selection theory in biology. Taking the organismic concept of groups seriously therefore amounts to a revival of functionalism in the social sciences—not in its original form, which in part deserved its fate, but in a form that can be justified theoretically and verified empirically.

NOW AND THEN

To survey the social science literature on religion, it will help to begin with the present state of the art. Arguably the most dynamic current research program is that of Rodney Stark, William S. Bainbridge, and their colleagues, which is an admirable blend of sociology, anthropology, history, psychology, and economics (Stark and Bainbridge 1985, 1987, 1997; Stark 1996, 1999; Finke and Stark 1992; Stark and Finke 2000). Their work is also impressive for its interplay between theory and empirical research and for the diversity of its research methods, which range from ethnographies of modern-day cults, to nationwide and worldwide survey data, to the ingenious construction of quantitative data bases from historical material. When I first approached this literature, I felt very much like a person in a rowboat pulling alongside a fleet of battleships!

Stark and his colleagues study religion from the perspective of economics and rational choice theory. Religion is envisioned as an economic exchange between people and imagined supernatural agents for goods that are scarce (e.g., rain during a drought) or impossible (e.g., immortal life) to obtain in the real world. Religious belief is therefore rational in the sense of employing cost-benefit reasoning. In addition, many details of religious belief and practice can be predicted from an economic perspective. Stark's (1999) theory of religion is ambitious: "all aspects of religion—belief, emotion, ritual, prayer, sacrifice, mysticism, and miracle—can be understood on the basis of exchange relations between humans and supernatural beings" (264). I will hereafter refer to this as the rational choice theory of religion.

The paper from which this quotation was obtained is an update of a book-length theory of religion (Stark and Bainbridge 1987). Both versions have the virtue of presenting major propositions in the form of crisp statements accompanied by equally concise definitions of terms. This format allows me to summarize Stark's latest version of the rational choice theory of religion in his own words, as shown in table 2.1.

To facilitate comparison, the hypothesis that religious groups function as adaptive units is summarized as a list of propositions in table 2.2.[1] Groups consist of people who actually interact with each other with respect to the activity at hand, as discussed in chapter 1. Remarkably, Stark's list fails to address the issues at the heart of my list. The essence of Stark's theory is that goods can either be procured by human action, in which case religious belief is unnecessary (proposition 7), or goods cannot be procured by human action, in which case supernatural agents are invented to provide the unprovidable. We pray to God for everlasting life, not to convey us to work in the morning. As Buckser (1995) puts it in one of the quotations that begin this chapter, according to Stark and Bainbridge religion gives us what we cannot have. Missing entirely from this conception is the category of goods that can be procured by human action, but only by coordinated human action, and the role of religion in achieving the required coordination. In short, Stark's theory ignores the fundamental problem of social life and the role of religion in its solution.

In evolutionary terms, Stark's list of propositions would be classified as a byproduct, or "spandrel" explanation. Propositions 1, 2, and 3 describe psychological attributes that are highly adaptive in general and can easily be explained as a product of natural selection. However, their manifestation in the case of religion is not adaptive. If religion does not actually deliver the scarce resources that supernatural agents are invented to provide, the entire enterprise is a waste of time as far as survival and reproduction are concerned. A mutant human race with the ability to employ its psychological attributes only when they deliver worldly benefits, turning them off otherwise, would quickly replace *Homo religiosis*. Religion persists only because people cannot fine-tune their psychological attributes to this degree. Religion is on the cost side of a cost-benefit equation as far as the evolution of human psychology is concerned.

Throughout their writings, rational choice theorists tend to be highly critical of functionalism, portraying it as a dead tradition that by all means should remain buried. At times, Stark is so bitingly satirical that he seems to deny functionalists the capacity for rational thought that he grants to religious zealots. Thus, we seem to have a clear case of a byproduct theory

---

Table 2.1 Stark's (1999) rational choice theory of religion and its implications, stated as a list of propositions

---

1. Within the limits of their information and understanding, restricted by available options, guided by their preferences and tastes, humans attempt to make rational choices.

2. Humans are conscious beings having memory and intelligence, who are able to formulate *explanations* about how rewards can be gained and costs avoided.

3. Humans will attempt to *evaluate* explanations on the basis of results, *retaining* those that seem to work most efficiently.

4. Rewards are always limited in supply, including some that simply *do not exist* in the observable world.

5. To the degree that rewards are scarce, or are not directly available at all, humans will tend to formulate and accept explanations for obtaining the reward in the *distant future* or in some other *nonverifiable context.*

6. In pursuit of rewards, humans will seek to *utilize and manipulate the supernatural.*

7. Humans will *not* have recourse to the supernatural when a cheaper or more *efficient alternative is known and available.*

8. In pursuit of rewards, humans will seek to *exchange with a god or gods.*

9. The greater number of gods worshipped by a group, the *lower the price of exchanging with each.*

10. In exchanging with the gods, humans will pay higher prices to the extent that the gods are believed to be more *dependable.*

11. In exchanging with the gods, humans will pay higher prices to the extent that the gods are believed to be more *responsive.*

12. In exchanging with the gods, humans will pay higher prices to the extent that the gods are believed to be of *greater scope.*

13. The greater their scope (and the more responsive they are) the more plausible it will be that gods can provide *otherworldly rewards.* Conversely, exchanges with gods of smaller scope will tend to be limited to *worldly rewards.*

14. In pursuit of otherworldly rewards, humans will accept an *extended* exchange relationship.

15. In pursuit of otherworldly rewards, humans will accept an *exclusive* exchange relationship.

16. People will seek to *delay* their *payment* of religious costs.

17. People will seek to *minimize* their religious costs.

18. A religious organization will be able to *require extended and exclusive commitments* to the extent that it *offers otherworldly rewards.*

19. Magic cannot generate extended or exclusive patterns of exchange.

20. Magicians will serve individual clients, not lead an organization.

Table 2.1 (*Continued*)

21. Otherworldly rewards entail risk.
22. An individual's *confidence* in religious explanations concerning otherworldly rewards is strengthened to the extent that *others express* their confidence in them.
23. *Confidence* in religious explanations concerning otherworldly rewards is strengthened to the extent that people participate in *religious rituals.*
24. *Prayer* builds bonds of affection and confidence between humans and a god or gods.
25. *Confidence* in explanations offered by religion concerning otherworldly rewards will increase to the degree that *miracles* are credited to the religion.
26. *Confidence* in explanations offered by religion concerning otherworldly rewards will increase to the degree that people have *mystical experiences.*
27. Vigorous *efforts* by religious *organizations* are required to *motivate and sustain* high levels of religious *commitment.*

Table 2.2 The hypothesis that religious groups function as adaptive units, stated as a list of propositions

1. Many resources and other valuable commodities can be achieved only by the *coordinated action* of individuals. When members of a social group act in such a fashion, they function as an *adaptive unit.*
2. Group-level adaptations do not easily evolve by natural selection because pro-social behaviors do not automatically increase relative fitness within groups. A process of *group selection* is required whereby well-functioning groups out-compete other groups.
3. Human groups function as adaptive units primarily by having a *moral system* that regulates behavior within the group.
4. Moral systems are frequently expressed in religious terms. Many features of religion, such as the nature of supernatural agents and their relationships with humans, can be explained as adaptations designed to enable human groups to function as adaptive units.
5. In some cases, adaptive features of religion evolve by an ongoing process of blind variation and selective retention; there are many social experiments, a few of which succeed.
6. In other cases, adaptive features of religion are the direct product of psychological processes such as conscious design or imitation. However, these psychological processes must themselves be understood as the product of multi-level selection operating in the more distant past.

of religion that has triumphed over a theory based on group-level adaptation. However, this impression is superficial. Before I attempt a closer analysis, we need to acquaint ourselves with the tradition of functionalism that appears to have failed so miserably.

DURKHEIM REVISITED

My main exemplar of functionalism will be Emile Durkheim's *Elementary Forms of Religious Life* ([1912] 1995). Prior to Durkheim, the two most influential theories of religion were known as "animism" and "naturism." According to animism, spiritual belief originated from the experience of dreaming, in which a phantom version of oneself appears capable of leaving the body and traveling long distances. Sleep, fainting, madness, and death all lead to the notion of a world of spirits who enter and leave human bodies at will. Once this world is imagined, it can explain anything:

> In this way they constitute a veritable arsenal of causes, always at hand, never leaving the mind that is in search of explanations unequipped. Does a man seem inspired; does he speak with eloquence; does he seem lifted above both himself and the ordinary level of men? It is because a benevolent spirit is in him, animating him. Is another man taken by a seizure or by madness? An evil spirit has entered his body, agitating him. There is no sickness that cannot be put down to some such influence. In this way, the power of souls increases from all that is attributed to them, so much so that, in the end, man finds himself a captive in this imaginary world, even though he is its creator and model. He becomes the vassal of those spiritual forces that he has made with his own hands and in his own image. For if these souls are so much in control of health and illness and of good and evil things, it is wise to seek their benevolence or to appease them when they are annoyed. From thence come offering, sacrifices, prayers—in short, the whole apparatus of religious observances. (Durkheim [1912] 1995, 49)

Naturism also portrays the religious believer as a captive in an imaginary world of his own making, but by a different pathway: awe of the forces of nature rather than the experience of dreaming. In modern evolutionary terms, both animism and naturism would be called byproduct theories, similar in spirit, if not in detail, to Stark's list of propositions. The human capacity for thought is broadly adaptive, but its particular manifestation in the case of religion has no function and can be costly to the extent that it misrepresents the world and leads to inappropriate behaviors. The passage quoted above even anticipates the modern concept of "selfish memes" (Dawkins 1976; Boyer 1994, 2001; Blackmore 1999), which envisions culture as a parasitic organism in its own right that exploits its human host.

Durkheim doubted that something as pervasive and influential as religion could be so dysfunctional. Early humankind lived too close to the edge of survival for such idle theorizing. Beliefs that failed to deliver practical benefits would soon be discarded in favor of more adaptive beliefs.

> That man has an interest in knowing the world around him and that, consequently, his reflection was quickly applied to it, everyone will readily accept. The help of the things with which he was in immediate contact was so necessary that he inevitably tried to investigate their nature. But if, as Naturism contends, religious thought was born from these particular reflections, then it becomes inexplicable that religious thought should have survived the first tests made, and unintelligible that religious thought has been maintained. If, in fact, we have a need to know things, it is in order to act in a manner appropriate to them. But the representation of the universe that religion gives us, especially at the beginning, is too grossly incomplete to have been able to bring about practices that had secular utility. According to that representation of the universe, things are nothing less than living, thinking beings—consciousnesses and personalities like those the religious imagination has made into the agents of cosmic phenomena. So it is not by conceiving them in that form and treating them according to that notion that man could have made them helpful to him. It is not by praying to them, celebrating them in feasts and sacrifices, and imposing fasts and privations on himself that he could have prevented them from harming him or obliged them to serve his purposes. Such procedures could have succeeded only on very rare occasions—miraculously, so to speak. If the point of religion was to give us a representation of the world that would guide us in our dealings with it, then religion was in no position to carry out its function, and humanity would not have been slow to notice that fact: Failures, infinitely more common than successes, would have notified them very quickly that they were on the wrong path;

and religion, constantly shaken by these constant disappointments, would have been unable to last. (Durkheim [1912] 1995, 76–77)

Since religious belief is such a poor representation of the natural world, its "secular utility" must reside elsewhere. Durkheim proposed that religion functions as an organizer of social life, both by defining groups and by prescribing the behaviors of its members. For Durkheim, the essence of religion was a distinction between the sacred and the profane: "A religion is a unified system of beliefs and practices relative to sacred things, that is to say, things set apart and forbidden—beliefs and practices which unite into one single moral community called a Church, all those who adhere to them" (44). The similarity between this passage and the biological concept of human groups unified by moral systems that we arrived at in chapter 1 is unmistakable. In modern evolutionary terms, Durkheim interpreted religion as an adaptation that enables human groups to function as harmonious and coordinated units. I take this to be the central thesis of functionalism in the social sciences as it relates to religion.

In addition to framing the central thesis, Durkheim also had a complex and sophisticated vision of exactly how religion performs its functions. He thought that a social group is such an abstract entity that it needs to be represented by a set of symbols to be comprehended by the human mind: "In all its aspects and at every moment of history, social life is only possible thanks to a vast symbolism" (233). Religion is therefore a symbolic representation of society. For example, human tribes are often divided into a number of clans whose membership is based on a complex mix of affinal and genetic relationships. Each clan is represented by a totem (usually an animal or a plant) that identifies clan membership, and associated with each totem are a variety of sacred objects and practices that guide the behavior of clan members. Durkheim felt that the symbolic badge of group membership and the aura of sacredness surrounding prescribed behaviors were required for clans to exist as functioning groups. He also felt that periodic gatherings were required to maintain the integrity of groups. The religious rituals and other festivities held during these gatherings were so emotionally intense that they gave force to group identity when its members were dispersed.

Some of Durkheim's specific proposals still sound plausible today. For example, Deacon (1998) has recently argued that symbolic thought sets humans apart from all other animals and evolved to enable enforced social contracts such as marriage. If he is correct, it will be an impressive confirmation of Durkheim's claim that human social life is only possible thanks to a vast symbolism. However, some of Durkheim's other proposals sound antiquated today; this is hardly surprising, since much has happened in the social sciences since 1912! For our purposes we need to distinguish Durkheim's general thesis that religion is a group-level adaptation from his many specific proposals about how religion performs its various functions. Was Durkheim right about the central thesis but wrong about some of the details, or was he wrong about the central thesis?

THE NEXT GENERATION

Stark and other rational choice theorists discuss Durkheim in a number of contexts. Stark thinks that Durkheim defined religion too broadly (it should be restricted to belief in supernatural agents) but credits him for correctly distinguishing between religion and magic. In general, however, Stark regards Durkheim's work in particular and functionalism in general as a paradigm that was rejected long ago and that does not need revisiting, citing the great British anthropologist E. E. Evans-Pritchard (1956, 313), who wrote: "It was Durkheim, not the savage, who made society into a god."

Unlike Durkheim, Evans-Pritchard lived among the people he studied. Not only are his ethnographies of the Nuer and other African tribes still regarded as classics, but he also wrote a general book on theories of primitive religion (Evans-Pritchard 1965), which makes him a good exemplar of the generation of anthropologists that followed Durkheim.

Evans-Pritchard (1965, 56–64) provides a concise and accurate summary of Durkheim's theory of religion that culminates in the following assessment:

> Durkheim's thesis is more than just neat; it is brilliant and imaginative, almost poetical; and he had an insight into a psychological fundamental of religion: the elimination of the self, the denial of individuality, its having no meaning, or even existence, save as part of something greater, and other, than the self. But I am afraid that we must once more say that it is a just-so story. (64)

Evans-Pritchard then weighed in on the details of Durkheim's thesis. As a hardened field anthropologist, he was more interested in the empirical details than in the logic and theory. He did not see a rigid dichotomy between the sacred and profane. When Zande shrines were not in ritual use they served as convenient props for spears. Among the Australian aboriginals whom Durkheim analyzed, it was the co-residential groups and tribes, not the clans identified by separate totems, that functioned as corporate units. Furthermore, Australian totemism provided a poor foundation for a general theory of religion. One can almost hear Evans-Pritchard heaving a sigh of exasperation as he ends his litany with the following statement: "if only Tylor, Marett, Durkheim, and all the rest of them could have spent a few weeks among the peoples about whom they so freely wrote" (67).

Although some of Durkheim's specific proposals were dismantled by Evans-Pritchard and others of his generation, it would be wrong to conclude that the general thesis of functionalism was similarly dismantled. The passage quoted above singles out the elimination of the self, save as part of something greater, as a psychological fundamental of religion. Evans-Pritchard's own books are peppered with references to groups as "corporate units." He might have disagreed with Durkheim on the status of clans as corporate units, but not on the basic existence of groups as corporate units or the role of religion in structuring such groups. Indeed, he is perhaps best known for his concept of segmentation: the organization of leaderless tribes into a nested hierarchy of groups that can become functionally organized at any level, depending on the scale of the environmental challenge (usually warfare).

Evans-Pritchard's (1956) analysis of Nuer religion is fully consistent with the general thesis of functionalism and the biological concept of human groups that we arrived at in chapter 1. In fact, one of his most arresting observations is that Nuer religion is similar to the Judaism of the Old Testament—not because of any historical connection but because both were derived from herding cultures whose lives were dominated by their livestock and their own social affairs. If true, this would be an example of convergent cultural evolution that would be difficult to explain without invoking the concepts of adaptation and natural selection.

Table 2.3 lists a number of passages from Evans-Pritchard (1956) that interpret Nuer religion as integral to the functional organization of Nuer society. I hope I will be forgiven for overkill, but it is important to establish that when someone as authoritative as Evans-Pritchard criticized Durkheim, he was not rejecting the interpretation of human groups as adaptive units, the role of religion in structuring society, or even the importance of ritual and symbolism provided by religion. What is true for Evans-Pritchard is also true for the entire generation of anthropologists who followed Durkheim and who account for most of the empirical knowledge

Table 2.3 Passages from E. E. Evans-Pritchard (1956) that interpret Nuer religion in terms of group-level functional organization

1. On humility and the similarity between Nuer religion and the Judaism of the Old Testament:

    In speaking about themselves as being like ants and as being simple Nuer show a humbleness in respect to God which contrasts with their proud, almost provocative, and toward strangers even insulting, bearing to men; and indeed humbleness . . . is a further element of meaning in the word *doar*, as is also humility, not contending against God but suffering without complaint. . . . I cannot convey the Nuer attitude better than by quoting the Book of Job: 'the Lord gave, and the Lord hath taken away; blessed be the name of the Lord.' (13)

2. On the concept of right and wrong in relation to God:

    This brings me to an extremely important Nuer concept, an understanding of which is very necessary to a correct appreciation of their religious thought and practice. This is the concept of *cuong*. This word can mean 'upright' in the sense of standing, as, for example, in reference to the support of byres. . . . It is most commonly employed, however, with the meaning of 'in the right' in both a forensic and moral sense. The discussion in what we would call legal cases is for the purpose of determining who has the *cuong*, the right, in the case, or who has the most right; and in any argument about conduct the issue is always whether a person has conformed to the accepted norms of social life, for, if he has, then he as *cuong*, he has right on his side. We are concerned with the concept here both because it relates directly to man's behaviour towards God and other spiritual beings and the ghosts and because it relates to God in a more indirect way, in that he is regarded as the founder and guardian of morality. (16)

3. On the consequences of right and wrong conduct:

    I do not want to suggest that God is thought to be an immediate sanction for all conduct, but I must emphasize that the Nuer are of one voice in saying that sooner or later and in one way or another good will follow right conduct and ill will follow wrong conduct. People may not reap their rewards for good acts and punishments for bad acts for a long time, but the consequences of both follow behind (*gwor*) them and in the end catch up on those responsible for them. You give milk to a man when he has no lactating cows, or meat and fish to him when he is hungry, or you befriend him in other ways, though he is no close kinsman of yours. He blesses you, saying that your age-mates will die while your children grow old with you. God will see your charity and give you long life. Those who have lived among the Nuer must have heard, and received, their blessings. (16–17)

Table 2.3 (*Continued*)

> The Nuer have the idea that if a man keeps in the right—does
> not break divinely sanctioned interdictions, does not wrong others,
> and fulfills his obligations to spiritual beings and the ghosts and
> to his kith and kin—he will avoid, not all misfortunes, for some
> misfortunes come to one and all alike, but those extra and special
> misfortunes which come from *dueri*, faults, and are to be regarded as
> castigations. . . . Not only a sin (a breach of certain interdictions) but
> also any wrong conduct to persons is spoken of as a *duer*, a fault. Any
> failure to conform to the accepted norms of behaviour toward a
> member of one's family, kin, age-set, a guest, and so forth is a fault
> which may bring about evil consequences through either an ex-
> pressed curse or a silent curse contained in anger and resentment,
> though the misfortunes which follow are regarded by Nuer as coming
> ultimately from God, who supports the cause of the man who has
> the *cuong*, the right in the matter, and punishes the person who is at
> fault (*dwir*), for it is God alone who makes the curse operative. Nuer
> are quite explicit on this point. What, then, Nuer ideas on the matter
> amount to is, in our way of putting it, that if a man wishes to be
> in the right with God he must be in the right with men, that is, he
> must subordinate his interests as an individual to the moral order of
> society. (18)

4. On extending morality to the entire community:

> It is of course natural as well as noticeable that close kin stick to-
> gether in danger and when a wrong has been done to any one of them,
> but Nuer also quite clearly show that they feel that a misfortune for
> any member of their community is a misfortune for all, that when
> one suffers all suffer, and that if each is to be at peace all must be at
> peace. (24)

5. On mutual criticism:

> All kinsmen who can do so should attend a ceremony of this kind,
> especially close agnatic kinsmen, for this is an occasion when kin are
> expected to reveal any resentment they may have in their hearts
> towards one another. Each tells the others where they have been at
> fault (*duer*) in speech or act during the past year, and the issues
> raised are then settled peaceably by discussion. A man must not keep a
> grievance hidden, and if he does not reveal it now he must for ever
> keep silence. Nuer say that all the evil in their hearts is then blown
> away. I think that this is the meaning of the throwing of beer into the
> air. (43)

Table 2.3 (*Continued*)

Consequently, such moral faults as meanness, disloyalty, dishonesty, slander, lack of deference to seniors, and so forth, cannot be entirely dissociated from sin, for God may punish them even if those who have suffered from them take no action of their own account. Nuer seem to regard moral faults as accumulating and creating a condition of the person predisposing him to disaster, may then fall upon him on account of some act or omission which might not otherwise and by itself have brought it about. This is further suggested by the custom of confession at certain sacrifices, when it is necessary to reveal all resentments and grievances a man may have in his heart toward others. (193)

6. On the role of symbolism in defining corporate units:

Each of the totemic spirits is Spirit in a particular relationship to a lineage, which expresses its relation as an exclusive social group to God in the totemic refraction by respecting the creature which stands as a material symbol of the refraction. (91)

It is, I believe, in some such sense of the clan militant that we should interpret the spear symbolism of the clans. Nuer always speak of 'spear' where I have used the expression 'spear-name.' However, not only are there in almost all cases no actual spears, but it is not even the idea of any particular spear that they have in mind. What they have in mind is the clan as a whole for which the spear stands as a symbol not just as a spear but as representing the collective strength of the clan in its most conspicuous corporate activity, for clans and lineages are most easily and distinctly thought of as collectives—through their identification with tribes and tribal segments—in relation to war. (246)

7. On context sensitivity:

The great variety of meanings attached to the word *kwoth* [spirit] in different contexts and the manner in which Nuer pass, even in the same ceremony, from one to another may bewilder us. Nuer are not confused, because the difficulties which perplex us do not arise on the level of experience but only when an attempt is made to analyse and systematize Nuer religious thought. Nuer themselves do not feel the need to do this. Indeed, I myself never experienced when living with the Nuer and thinking in their words and categories any difficulty commensurate with that which confronts me now when I have to translate and interpret them. I suppose I moved from representation to representation, and backwards and forwards between the general and particular, much as Nuer do and without feeling that there was any lack of co-ordination in my thoughts or that any special effort to understand was required. It is when one tries to relate Nuer religious conceptions to one another by abstract analysis that the difficulties arise. (106)

Table 2.3 (*Continued*)

8. On death and the afterlife:

> Nuer avoid so far as possible speaking of death and when they have to do so they speak about it in such a way as to leave no doubt that they regard it as the most dreadful of all dreadful things. This horror of death fits in with their almost total lack of eschatology. Theirs is a this-worldly religion, a religion of abundant life and the fullness of days, and they neither pretend to know, nor, I think, do they care, what happens to them after death. (154)

9. On the importance of sincerity and truth in the presence of God:

> Nuer say that a man must make invocation in truth (*thuogh*). Even in recounting the most trivial details in a history of events which have led up to a situation in which sacrifice is necessary every statement made in the presence of God must be true; and I think this, rather than just a desire to interrupt or dispute, accounts for the emendations, additions, and contradictions of the assistants. . . . If the sacrifice is to be efficacious what is said must be true. (211)

> I have suggested that the fighting-spear has a symbolic meaning for Nuer besides what it means for them as a weapon and tool—that it is a projection of the self and stands for the self. This is most important for an understanding of Nuer sacrifice. Its manipulation is most common, and for a study of religion the most significant, sacrifices, the personal and piacular ones, expresses, if our interpretation is right, the throwing of the whole person into the intention of the sacrifice. It is not only said but also thought, desired and felt. Not only the lips make it but also the mind, the will, and the heart. (239)

10. On the conflict resolution function of one type of priest:

> Lineages of leopard-skin priests are found in all tribal sections, and in most parts of Nuerland they are in the category of *rul*, strangers, and not of *diel*, members of the clans which own the tribal territories. It is necessary that they should be widely spread, because their services are essential to Nuer everywhere, and it is significant that they are generally not members of lineages identified with political groups, because they have to act as peacemakers between such groups. . . . They are like Levi, divided by Jacob and scattered in Israel. (292)

> The leopard-skin priest for the most part performs sacrifices in situations in which two groups are opposed to one another and which therefore require a person unidentified by lineage attachment to either to act on behalf of the whole community. The priest thus has a central position in the social structure rather than in religious thought, for his priestly functions are exclusive not because, on account of the sanctity

---

Table 2.3 (*Continued*)

---

of his office, only he can perform sacrifices, but because representatives
of neither party to a dispute can effectively act in the circumstances obtaining. As he entirely lacks any real political authority or powers it is
understandable that he could not carry out his functions unless during
their performance his person was sacrosanct. (300)

11.   On the internalization of religion:

Though prayer and sacrifice are exterior actions, Nuer religion is
ultimately an interior state. This state is externalized in rites which we
can observe, but their meaning depends finally on an awareness of God
and that men are dependent on him and must be resigned to his will.
(322)

---

of so-called primitive people in something close to their pristine state. To
choose another example, Victor Turner ([1969] 1995) analyzed religious
ritual in terms of two key concepts: communitas and structure. Structure is
the system of roles related to age, sex, and status that people in a community
occupy. Communitas is a conception of the community as an egalitarian
unit in which all members, from highest to lowest, have a moral claim. The
purpose of structure is to implement the spirit of communitas. As far as
social norms are concerned, prescribed roles are intended to serve the interests of the community and should not be exploited for personal gain.

> An incumbent of high status is peculiarly tempted to use the authority
> vested in him by society to satisfy these private and privative wishes. But
> he should regard his privileges as gifts of the whole community, which in
> the final issue has an overright over all his actions. Structure and the high
> offices provided by structure are thus seen as instrumentalities of the
> commonweal, not as means of personal aggrandizement. (Turner [1965]
> 1995, 104)

Of course individuals often do abuse the power invested in them by their
communities. As we saw in chapter 1, human groups achieve their functional organization not entirely by self-restraint (although this can be an
important factor) but by mutual vigilance and social control. According
to Turner, ritual is one important mechanism that keeps structure and
communitas bound to each other. A common feature of ritual in both
traditional and modern societies involves the stripping away of status,

especially during the transition from one social role to another. One mem-
orable example from Gabon involves the election of a new king, who is
chosen secretly by the village elders and is kept ignorant of his fate until
the following event occurs:

> It happened that Njogoni, a good friend of my own, was elected. The
> choice fell on him, in part because he came of a good family, but chiefly
> because he was a favourite of the people and could get the most votes. I
> do not think that Njogoni had the slightest suspicion of his elevation. As
> he was walking on the shore on the morning of the seventh day [after the
> death of the former king] he was suddenly set upon by the entire popu-
> lace, who proceeded to a ceremony which is preliminary to the crowning
> and must deter any but the most ambitious man from aspiring to the
> crown. They surrounded him in a dense crowd, and then began to heap
> upon him every manner of abuse that the worst of mobs could imagine.
> Some spat in his face; some beat him with their fists; some kicked him;
> others threw disgusting objects at him; while those unlucky ones who
> stood on the outside, and could reach the poor fellow only with their
> voices, assiduously cursed him, his father, his mother, his sisters and broth-
> ers, and all his ancestors to the remotest generation. A stranger would not
> have given a cent for the life of him who was presently to be crowned.
>
> Amid all the noise and struggle, I caught the words which explained
> all this to me; for every few minutes some fellow, administering a spe-
> cially severe blow or kick, would shout out, "You are not our king yet; for
> a little while we will do what we please with you. By-and-by we will have
> to do your will."
>
> Njogoni bore himself like a man and prospective King. He kept his
> temper, and took all the abuse with a smiling face. When it had lasted
> about half an hour they took him to the house of the old king. Here he
> was seated, and became again for a little while the victim of his people's
> curses.
>
> Then all became silent; and the elders of the people rose and said, sol-
> emnly (the people repeating after them), "now we choose you for our
> king; we engage to listen to you and to obey you." He was then dressed
> in a red gown, and received the greatest marks of respect from all who
> had just now abused him. (from Du Chaillu 1868; quoted in Turner
> [1965] 1995, 171)

This passage wonderfully illustrates the spirit of egalitarianism that pervades human communities long after they have become structurally hier-

archical. Turner does not comment or provide evidence for the effectiveness of such humbling rituals in curtailing selfish behavior, which should be an important objective for future research. However, the meaning and intention of such rituals, as interpreted by Turner, is highly consistent with the conception of human groups unified by moral systems that we arrived at in chapter 1.

Until recently, the only way to learn about the customs of so-called primitive people was through the writing of Western anthropologists. Increasingly, however, members of these cultures are speaking for themselves. An interesting example is Malidoma Patrice Somé, a member of the West African Dagara culture, who has attempted to interpret the nature of ritual for an American audience (Somé 1997). According to Some, rituals and social responsibilities are "inseparable" (11). Table 2.4 lists representative passages that gratifyingly support the central thesis of Durkheim, Evans-Pritchard, Turner, and many other anthropologists of their generations. The connection between religion and functionally organized society is not just a figment of Durkheim's or of the Western imagination.

Table 2.4 Passages from Somé (1997) that interpret Dagara ritual in terms of group-level functional organization

---

1. On ritual, the spirit world, and community:

   For the Dagara, ritual is, above all else, the yardstick by which people measure their state of connection with the hidden ancestral realm, with which the entire community is genetically connected. In a way, the Dagara think of themselves as a projection of the spirit world. (12)

   Remember, the value of a ritual community is that it creates power that protects and helps all within the community. (64)

2. On ritual and social health:

   I am tempted to think that when the focus of everyday living displaces ritual in a given society, social decay begins to work from the inside out. The fading and disappearance of ritual in modern culture is, from the viewpoint of the Dagara, expressed in several ways: the weakening of links with the spirit world, and general alienation of people from themselves and others. In a context like this there are no elders to help anyone remember through initiation of his or her important place in the community. (14)

   A true community does not need a police force. The very presence of a law enforcement system in a community is an indication that something is not working. (50)

Table 2.4 (*Continued*)

3. On the consequences of selfishness and wrong-doing:

   The Dagara people, on the other hand, are suspicious of abundance. It translates into a cultural attitude that a person of abundance is a person too worldly to deal with hardship. This is an obvious trick from a god to put someone to sleep before the final blow. In my village there was a man whom everyone knew to be very poor, or at least he did not have anything that anyone would desire. His compound was as normal as anyone else's. He was married with several children. But one day that man began to act most strangely. First, there was delivery of a half-dozen cows from nearby Ghana by someone no one knew of. The man soon began to build a modern home and bought himself a brand new English bicycle. He wore clean clothes. He had changed in ways that were suspicious. And he showed abundance in ways that insulted the entire setup of the tribe. People waited for the inevitable. It occurred quickly. (15)

   Elders say that the real police in the village is Spirit that sees everybody. To do wrong is to insult the spirit realm. Whoever does this is punished immediately by Spirit, as in the case of a newly wed girl who stole grain from the granary of her father-in-law. She sold the millet grains in the supermarket and then went into the bush to look for some dry wood to bring back on her way home. She found a dry tree and began to cut it down. It was not long before she cut herself deep in the arm. She arrived home bleeding and screaming. Her father-in-law rushed to the diviner. These kinds of accidents don't happen without a reason. The diviner revealed that his daughter-in-law was a thief who was purging her evil action. Because she had transgressed the Spirit, she had invited an accident. The diviner added that the wound would not stop bleeding until she publicly admitted to the theft. It brought shame to both the father-in-law and the daughter. Her pleading guilty stopped the bleeding and allowed the wound to be dressed properly. (50–51)

4. On humility and the non-egoistic nature of ritual:

   The success of a ritual depends on the purpose of the individuals involved with it. Any ritual designed to satisfy an ego is a ritual for show, and therefore is a spiritual farce. But any ritual in which the persons involved invite the spirits to come and help in something that humans are not capable of handling by themselves or in which humans honor gift-giving from the divine, there is a likelihood of the ritual working. (27)

   Invocation is a call placed by a person to a spirit. To invoke the spirit is to call upon the invisible. The language of invocation must not

Table 2.4 (*Continued*)

be confused with order and command. . . . For anything to happen, the ritual must be dominated by humility. (53)

5. On mutual criticism:

Consequently, a person's social failures are brought out in the course of funeral ritual, and as a result create a special kind of grief. Death reminds the person who is not paying his or her social dues to the community that he or she must repent and grieve for past failures. (80)

6. On the distinction between the sacred and profane:

The profane is the allergy of the spirit. Within a ritual space, anything that is not sacred threatens to desecrate the hallowedness of what is happening. Consequently, it drives the spirit "crazy," and he who is in charge of the ritual is in big trouble. But when the space is kept clean, ritual yields great power to those involved in it. (39)

7. On the after-effects of ritual:

Whatever happens in a ritual space, some kind of power is released if given a freedom in which to live. This is the only way those who participate in the ritual can continue to benefit from the power. The forces aroused in the ritual function like a power plant into which every individual is hooked. When one leaves the ritual space, the power of the ritual goes wherever the person goes. Only in ritual can the "here" follow you to the "there." (42)

8. On the consequences of violating the sacred:

When back in the village, I asked what was wrong with him. A village acquaintance replied laughingly that he managed to steal the shrine of the ancestors with the intention of selling it to a group of white people. And he also spoke the unspeakable. "Where he is, is the place that his actions naturally put him." I realized that, to the village, this person was no longer alive, no longer existed. No one was either sad or happy about him. He was not there. (43)

FUNCTIONALISM TODAY

So far I have shown that functionalism was not rejected by the anthropologists who followed Durkheim, including those who lived among the people they studied. When and why, then, was functionalism rejected? There is no single answer to this question, since the social sciences are a vast archipelago of disciplines that only partially communicate with each other. The prevailing wisdom about functionalism varies among disciplines, even for the study of religion (see Allen et al. 1998 for a recently published volume celebrating Durkheim's *Elementary Forms of Religious Life*, in contrast to the disparaging views of most rational choice theorists). This kind of fragmentation within the social sciences is regrettable, especially for those, such as myself, who are attempting to forge an even larger synthesis between the social sciences and biology.

The most comprehensive modern assessment of functionalism can be obtained from philosophers, who are better than most practicing scientists at synthesizing across disciplines. A useful selection of philosophical articles on functionalism and the closely related subject of holism has been compiled by Martin and McIntyre (1994). These articles raise a number of issues that need to be examined from a modern evolutionary perspective. The following discussion will help to revive functionalism throughout the social sciences and will pave the way to our study of religion.

Holism

The idea that the whole is somehow more than the sum of its parts is one of the most common and also one of the most vaguely articulated themes associated with functionalism. Durkheim was eager to create a science of sociology and was not prone to mysticism. He was happy to concede that the social organism could exist only within individual minds, but he nevertheless insisted that a sociological level of explanation existed that could not be reduced to individual behavior. The alternative to Durkheim's view became known as methodological individualism, described by Watkins ([1957] 1994, 442) as follows:

> According to this principle, the ultimate constituents of the social world are individual people who act more or less appropriately in the light of their dispositions and understanding of their situation. Every complex social situation, institution, or event is the result of a particular configuration of individuals, their dispositions, situations, beliefs, and physical resources and environment. There may be unfinished or halfway explanations of large-scale phenomena (say, inflation) in terms of other large-scale phenomena (say, full employment); but we shall not have arrived at rock-bottom explanations of such large-scale phenomena until we have deduced an account of them from statements about the dispositions, belief, resources and interrelations of individuals. . . . And just as mechanism is contrasted with the organicist idea of physical fields, so methodological individualism is contrasted with sociological holism or organicism.

Methodological individualism swept through the social sciences and eclipsed "organicism" at about the same time that the theory of individual selection swept through evolutionary biology, eclipsing multilevel selection theory. Nevertheless, these two forms of individualism are very different from each other, as I will show. In addition, it might surprise some readers to learn that the extreme reductionism expressed by Watkins has not withstood critical scrutiny, enabling philosopher Elliott Sober (1999) to state: "If there is now a received view among philosophers of mind and philosophers of biology about reductionism, it is that reductionism is mistaken."

The most relevant form of holism for our purposes is based on the concept of adaptation. Consider all the artificial selection experiments that have been conducted on fruit flies (*Drosophila*): they have been given short wings, long wings, no wings, many bristles, few bristles—the list goes on and on. The question "Why do these particular fruit flies have this particular phenotype?" has two answers. First, every phenotype is caused mechanistically by genes that interact with each other and their environment during development. Second, the phenotype exists because of a history of selection for that phenotype, coupled with heritable variation. These two explanations are usually labeled "proximate" and "ultimate" respectively, and there is a sense in which the latter is more fundamental than the former. As one classic example, Cohan (1984) divided a single population of fruit flies into a number of isolated smaller populations. He then selected the same trait (wing vein length) in each isolated population, examining the response to selection and the underlying genetic mechanisms in each case. It turned out that the same phenotypic trait of long wing veins evolved by different genetic mechanisms. A single ultimate-level explanation sufficed for all the populations, but different proximate-level explanations were required. More generally, the natural world is full of species that have evolved similar solutions to life's problems (e.g., hard exteriors as protection from predators), even though they are composed of different genes and physical materials (e.g., chitin for beetles, cellulose for plant seeds, and calcium carbonate for snails).

In short, heritable variation and selection provide a solid foundation for the holistic claim that the parts permit the properties of the whole but do not cause the properties of the whole.[2] A lump of clay permits but does not cause the form provided by the sculptor. To the extent that the physical make-up of organisms provides heritable variation, it becomes a malleable clay that can be sculpted by selection. Evolutionary biologists rely upon this kind of holism all the time. They confidently predict that desert animals should be sandy colored, a bird's wing should be aerodynamically efficient, and small fish should be timid in the presence of predators without any reference whatsoever to the genes or physical material that make up these organisms. Proximate explanation ("the sandy color in this species is caused by chromatin granules, which are coded by genes on the fourth chromosome") complements ultimate explanation ("the sandy color is caused by a history of selection favoring cryptic coloration") but never substitutes for it.

The distinction between proximate and ultimate explanation, which lies at the heart of evolutionary analysis and which will figure importantly in our study of religion, stands in contrast to Watkins's use of the word "ultimate" in the passage quoted above. For Watkins, the whole can only be truly understood in terms of the interactions among its parts, which are its "ultimate" constituents. This might be true for objects that have not been molded by natural selection. If you are handed a mineral and asked to explain its properties, what else can you do but look at its parts and their interactions? However, the molding action of natural selection endows the word "ultimate" with a new meaning that gives the whole priority over its clay-like parts. That is why the theory of individual selection, which relies heavily on the concept of adaptation, is different from methodological individualism, which relies heavily on the concept of mechanistic reductionism (Wilson 1988).

So far I have used individual organisms as the wholes in my discussion of holism. I have shown that Watkins's negative comments about "organicism" are mistaken when applied to creatures such as an insect, a fish, or a person. Multilevel selection theory allows us to frame-shift the entire discussion upward, in which the wholes are social groups and the parts are their individual members. To the degree that groups are units of selection, they possess properties that are permitted but not caused by their individual members. This is exactly what Durkheim and other functionalists were reaching for, which the tradition of methodological individualism erroneously seemed to deny. A position in the social sciences that in the past appeared wooly-minded and mystical can be placed on a rock-solid evolutionary foundation.

Complexity

Adaptationism is the most relevant form of holism/functionalism for the purposes of this book. Another form of holism/functionalism, based on complex interactions, needs to be distinguished and set aside even though it is interesting and important in its own right.[3] Consider the famous examples of water and salt, whose properties are difficult to predict from their parts (hydrogen and oxygen in the case of water and sodium and chloride in the case of salt). This concept of holism is based on complex interactions rather than functional organization. Salt and water have no purpose; their properties simply reside mostly in the interactions among their parts rather than in the properties of the parts as isolated units. In the same way, functionalism in the social sciences often seems to stress complexity and interconnectedness rather than functional organization per se (e.g., Malinowski 1944, 158–59). One example is a book called *Dynamic Functionalism* (Faia 1986), which stresses complex feedback processes and multiple stable equilibria ("absorbing Markov chains"). Along with Kincaid (in Martin and McIntyre 1994, 417), I think that this rendering of functionalism "casts the net too widely." I take it as a given that human cultures and social interactions are tremendously complex and interconnected, like water and salt only more so. Unfortunately, dysfunction can be as complex and locally stable as function, so the term "functionalism" should be avoided for concepts of holism based on complexity and restricted to the direct or indirect products of natural selection.

Functionalism and multilevel selection

Virtually everyone who has written on functionalism agrees that it is a legitimate form of explanation for individual-level adaptations that evolved by natural selection. There is nothing wrong with saying that the heart functions to circulate blood, that the turtle's shell functions to protect it from predators, etc., as long as these traits really did evolve by natural selection for the reasons that are ascribed to them. The problem with functionalism begins when we attempt to explain the properties of groups in general and especially human groups, for which the influence of natural selection is not so obvious. The rejection of group selection by evolutionary biologists and hostility toward evolutionary approaches to human behavior by many social scientists combined over the years to restrict functional explanation to the individual properties of nonhuman species.

I fully agree that functional explanations must be judged by the same criteria in evolutionary biology and in the social sciences. What has changed is the likelihood that the criteria are satisfied for the properties of groups in general and human groups in particular. Since philosophers and social scientists already acknowledge the legitimacy of functionalism at the individual level in nonhuman species when warranted by natural selection, the developments in evolutionary biology that I outlined in chapter 1 automatically set the stage for the return of group-level functionalism in the social sciences.

Just-so stories

A common criticism of adaptationist hypotheses in biology and functionalist hypotheses in the social sciences is that they are difficult to test. The derogatory term "just-so story," an allusion to the fanciful tales of Rudyard Kipling, was used by Evans-Pritchard against Durkheim (see the passage quoted above) long before Gould and Lewontin (1979) used it against what they regarded as adaptationist excesses in biology.

Of course, no one would dream of saying outright "this hypothesis is false because it is difficult to test." Science is a labor-intensive process, and some of the most important nuts are also the most difficult to crack. The "just-so story" criticism of functionalism implicitly assumes that nonfunctionalist hypotheses are easier to test and can substitute for functionalist hypotheses. Neither of these assumptions is likely to be true. I have already discussed the fact that proximate and ultimate explanations do not substitute for each other. In addition, adaptationism enjoys its status in evolutionary biology in part because it is *simpler* to employ than nonfunctionalist approaches. Often only a little knowledge suffices to make an initial prediction about the properties of organisms that enhance fitness in their environments (e.g., fish in streams with predators should be more timid than fish in streams without predators). In contrast, predictions based on phylogeny, genetics, development, and physiology require far more effort. As with predictions, so also with testing. Why should the hypothesis that fish are timid in the presence of predators be more difficult to test than, say, the hypothesis that fish development constrains heritable variation? Mature research programs in biology pay equal attention to ultimate and proximate explanation, but they often begin with an adaptationist hypothesis that provides the best and most economical "first guess" about the properties of the organism (see Hempel 1959, reprinted in Martin and McIntyre 1994, 371, for a similar view of functionalism in the social sciences). Indeed, discrepancies between adaptationist predictions and the properties of real organisms often lead to the discovery of nonadaptive factors that would have been missed otherwise. If we discover that fish are not timid in the presence of predators, despite the fact that it would increase their fitness, we might be tempted to look for factors that constrain adaptation and natural selection.

I do not mean to underestimate the challenge of testing functionalist hypotheses in evolutionary biology or in the social sciences. However, it is equally excessive to regard functionalism in general as a giant compendium of just-so stories that are somehow immune to scientific inquiry. The fact is that with enough hard work, evolutionary biologists are able to demonstrate the presence or absence of adaptations beyond all reasonable doubt. The quality of the science and the reliability of its conclusions are as good as for any other field of inquiry.

Because I intend to use standard evolutionary methods to study religion, it will help to show how they are successfully used to study adaptations in nonhuman species. When they aren't in pet stores and household aquaria, guppies (*Peocilia reticulata*) live in tropical streams that typically include dangerous fish predators in their downstream but not in their upstream portions. Since predators are a major source of mortality for guppies, we can predict that downstream and upstream populations become locally adapted to the presence and absence of predators, respectively. This hypothesis can be tested with at least three different sources of evidence, most usefully in combination with each other (reviewed by Endler 1986, 1995).

The first source of evidence is based on the argument from design. An object designed for a given purpose must have certain properties to achieve its purpose, which are unlikely to have arisen by chance. The more numerous, complex, and interlocking the design features, the more compelling the evidence for their designed nature. William Paley (1805) popularized the argument from design as evidence for God, but in fact it only provides evidence for a designing agent—the hand of God, a human engineer, alien visitors from another planet, or the process of natural selection. Thus, if we are studying organisms and if we exclude special creation and alien designers from other planets from consideration, we are left with design as a source of evidence of evolutionary adaptation. In the case of guppies, predator avoidance involves a suite of morphological, behavioral, and life history traits that would be difficult to explain in other ways. Endler (1995) provides a list of forty-seven traits that have been examined so far. One of these is age at first reproduction. In a downstream environment where every day might be a guppy's last, it is adaptive to have babies as soon as possible. In an upstream environment where the predators are so ineffective that only small guppies are in danger, it is adaptive to grow to a large size before starting to have babies. Upstream guppies are therefore predicted to have a later age of first reproduction than downstream guppies. When predictions such as this one are confirmed for a large suite of traits, it provides compelling evidence for adaptation and natural selection.

The argument from design can be based on the properties of a single organism or structure (e.g., the eye, or Paley's example of a watch, which is so elaborately constructed for a purpose that it couldn't exist by chance). The second source of evidence expands the view to include a comparison of organisms that occupy different environments. If natural selection adapts organisms to their environment, then the organisms should change as the environment changes (in more technical language, there should be a phenotype-environment correlation). Not only should downstream guppies have a suite of traits adapting them to the presence of predators, but upstream guppies should have another suite of traits adapting them to the absence of predators. Moreover, the geographical range of guppies includes dozens of river systems that have been separated from each other for millions of years. The river systems differ in their particular predator species, geochemistry, and other features, but all have the same downstream-upstream gradient of predator risk, providing a natural replicated experiment. In some cases the transition between upstream and downstream is amazingly abrupt, as when a waterfall acts as a barrier for the predator species but not for the guppies. In these cases, guppies that live only a few meters apart from each other (on each side of the waterfall) should differ in their phenotypic traits. If the upstream-downstream differences are found again and again in separate river systems, the adaptationist hypothesis receives powerful support. Subtle differences between river systems can also provide additional evidence. For example, in some streams the major predator is a crustacean rather than a fish. Crustaceans have a different visual system than vertebrates and are blind to the color orange, which gives their guppy prey a private wavelength for displaying to each other while remaining cryptic to their predators. It turns out that male guppies in the presence of fish predators are cryptic to the vertebrate eye, while male guppies in the presence of crustacean predators are ablaze with orange spots.

The third source of evidence involves tracking the evolutionary process. The ultimate proof of an adaptation is to actually watch it evolve. Natural selection was once assumed to be such a slow process that only the products, and not the process, could be directly observed. This view has been shattered by dozens of empirical studies, as eloquently described by Weiner (1994) in his Pulitzer prize-winning book *The beak of the finch*. In the case of guppies, downstream fish transplanted to upstream tributaries have evolved the upstream suite of traits within ten generations. To determine that predators were the causative agent, the same experiment was performed in the laboratory with identical physical environments that differed only in the presence and absence of predators. Other selection experiments altered the size and color of the gravel that forms the physical back-ground against which guppies are seen by their predators. Since avoiding predators requires matching one's background, changing the spots in the environment actually changed the spots on the guppies in only a few gen-erations (Endler 1986). Could there be a more convincing demonstration of organisms as a reflection of their environment?

When all three sources of evidence are taken together, any reasonable person must conclude that guppies are locally adapted to the presence and absence of predators in their environments. Even creationists have accepted this conclusion for evolutionary research on microevolution, which in part has caused them to shift their definition of "natural kind" to a higher taxonomic scale. This does not mean that guppies are perfectly adapted in every way or that constraints on the process of natural selection can be ignored. As one example, Houde and Hankes (1997) studied two populations in which females preferred orange-colored spots in males, but in only one of these populations had the males responded to sexual selec-tion by evolving orange spots. More work is required to determine the factors that seem to have constrained natural selection in this case.

It is gratifying that empirical research can achieve the middle ground between functionalism and its alternatives that so often is lost in polemical debates. In addition, evolutionary methods for studying adaptations in nonhuman species can be applied to the study of religion, as I will attempt to show in subsequent chapters. While certain kinds of research on hu-mans are off-limits for obvious and well-justified ethical reasons, other kinds of research can be performed more easily on humans than on other species. In addition, solid progress can be made on the basis of descriptive information and "natural experiments" that exist in abundance for reli-gions around the world and throughout history. Rather than complaining about the difficulty of testing functionalist hypotheses, we need to roll up our sleeves and start using our proven tools on the material at hand.

Intentional behavior

Human behavior is often goal oriented, and the implements people employ to achieve goals are correspondingly functional. If I lock myself out of my house and fetch a ladder to enter the upstairs window, the func-tion of the ladder is to enable me to reach the window. The function of the lock and key is to enable me to gain access to my house while keeping others out, even if I occasionally lock myself out as well. Human life is awash with this kind of functionality, which exists at the level of groups in addition to individuals. A business corporation might be intricately designed by its executives to maximize profits. The very word corporation, which is derived from the Latin word for "body," implies functional organization above the level of the individual.

In an influential analysis of modes of scientific explanation, Elster (1983) properly distinguishes causal (= proximate) explanation from functional (= ultimate) explanation. However, he then tries to make another distinction between functional explanation and intentional explanation. My efforts to discuss functionalism with social scientists frequently run aground on this distinction, which therefore requires our attention.

Let's begin with biology, which Elster agrees provides the foundation for functionalism. Consider a species of zooplankton with a peculiar spike on its head. Suppose we conduct sufficient research to establish that the spike is an antipredator adaptation. Elster presumably would agree that the function of the spike is to protect the individual from predators. Now consider another species of zooplankton that is developmentally flexible. No spike develops when predators are absent but the presence of predators is sensed chemically and triggers the development of the spike. Suppose we conduct sufficient research to establish that the developmental mechanism is also an adaptation. In fact, the second species is historically derived from the first species. Its developmental flexibility allows it to "have its cake and eat it too," by dispensing with the spike when it is not needed.

Many examples of adaptive flexibility (also called "phenotypic plasticity") have accumulated in the biological literature. Mentality is not required; bacteria and plants can be adaptively flexible. It is easy to show theoretically that flexibility is favored in some kinds of environments but not others (Wilson and Yoshimura 1994). Some species (including humans) might even consist of a mix of developmentally flexible and inflexible genotypes (Wilson et al. 1994). I want to claim that adaptive flexibility does not change the functional interpretation of phenotypes. The function of the spike in both zooplankton species is to protect the individual from predators. The species differ in the proximate mechanisms that evolved to produce the spike, but that difference does not alter the functional interpretation of the spike. Similarly, guppies cannot change their spots as individuals and therefore require the passage of generations to match their background. Flounders, octopi, and many other species can change their color as individuals and can match their background in a few moments. In both cases the function of matching one's background is to avoid predators.

My claim should be uncontroversial, but what is human intentionality but an elaborate mechanism of environmental assessment, built by natural selection, that culminates on balance in adaptive phenotypes? Allman's (1999) panoramic review of brain evolution illustrates my point by starting with the brain-like functions of bacteria:

> Some of the most basic features of brains can be found in bacteria because even the simplest motile organisms must solve the problem of locating resources and avoiding toxins in a variable environment. Strictly speaking, these unicellular organisms do not have nervous systems, but nevertheless they exhibit remarkably complex behavior: They sense their environment through a large number of receptors and store this elaborate sensory input in the form of brief memory traces. Moreover, they integrate the inputs from these multiple memory sensory channels to produce adaptive movements. The revolution in our understanding of genetic mechanisms has made it possible to determine how these brainlike processes work at a molecular level in bacteria. (Allman 1999, 3)

Real brains have the same function of transforming environmental information into adaptive phenotypes. Human brains may be unique in their ability to process symbolic information (as argued by Deacon 1998), but that does not alter their basic function or the functionality of the activities that are motivated by human thought. If a person knowingly fashions a spear to ward off predators, why on earth should we avoid the conclusion that the function of the spear is to ward off predators, just because intentional thought was involved?

Elster's distinction between functional and intentional explanation seems so strange that the reader may wonder if I am misrepresenting his position. Therefore, consider his own example of how to explain the profit-maximizing behavior of business firms (Elster 1983, 57–58). Evidently, the Chicago school of economists determined that firms were maximizing their profits but that the members of the firms were largely unaware of the specific practices responsible for their success. This finding led the economists to speculate that the successful practices had evolved by a process of selection; firms vary in ways unknown to their members, and the best outcompete the worst. Another possibility is that members of firms recognize and imitate success, even if they don't understand how the specific practices that they imitate lead to success. In this case, the best practices spread by imitation rather than by the actual survival and reproduction of firms. According to Elster, only the raw evolutionary process of differential survival and reproduction qualifies as a functional explanation. Even spread by imitation qualifies as an intentional explanation. This is like saying that a spear cannot be explained by its function if the spear-maker knows what he is making, or even if he blindly imitates another person who owns a spear and brings home more game.

The evolutionary distinction between ultimate and proximate explanation can help us avoid this nonsensical conclusion while retaining the distinction that I suspect Elster is trying to make. The first and foremost question we must ask is whether an object of study counts as functional: are its properties designed to accomplish a given effect? A spear is a spear is a spear, regardless of whether it is made by intentional thought, blind imitation, or an even blinder process of differential survival and reproduction. The next question is to determine the designing agent or process. Is it the hand of God, a selection process, or an organism that itself is a product of a selection process? It is remarkable how well the first question can be answered without knowledge about the answer to the second question. William Harvey discovered the function of the heart and circulatory system centuries before Darwin. He was wrong about the designing agent but right about the function of the heart. The Chicago school of economists discovered the profit-maximizing behavior of firms without needing to settle the issue of imitation vs. takeover. This is one reason to regard intentional thought as one of several possible proximate mechanisms that create function rather than as an alternative to functional explanation.

If the designing agent is an organism, a new set of questions comes to the fore. In the case of humans, we might especially want to know if function can be attributed to the mental processes that Elster classifies as intentional. Thus, Elster's distinction remains important but can be appreciated while avoiding the bizarre consequences of treating intentional and functional explanations as fundamentally different from each other.

Latent and manifest functions

Despite my disagreement with Elster, there is a sense in which he correctly interprets the functionalist debate. Functionalists were not simply pointing out the functions of intentional behaviors that were already manifestly obvious. They were claiming latent functions for practices and institutions at a grand scale that were not products of conscious intentional thought. Stated in evolutionary terms, Elster thinks that conscious intentional thought is the only proximate mechanism that makes human life functional. There are no latent functions beyond our awareness, only those of which we are already conscious. If this position is correct, then functionalism as an intellectual tradition would be dead.

However, this position is completely untenable, based on the evolutionary principles outlined in chapter 1. Functionality in human life can be attributed to at least three proximate mechanisms beyond the conscious intentional mechanisms emphasized by Elster: (1) psychological processes at the individual level that lie beneath conscious awareness; (2) group-level processes that individuals partake in without conscious knowledge of their roles; and (3) ongoing processes of cultural evolution.

With respect to individual psychology, Elster evidently regards all animal mentality and unconscious human mentality as simple forms of associative learning, which makes the possibility of unconscious intentions inconceivable to him:

> It follows from this argument that the notion of unconscious intentions
> is no more coherent than that of a squared circle. It does not follow, how-
> ever, that it is impossible to make sense of the notion of the uncon-
> scious, if we conceive of it strictly as a mechanism for climbing along a
> pleasure-gradient. It would be absurd to impute to the unconscious the ca-
> pacity for waiting, for making sacrifices, for acting according to rules, etc.,
> for these modes of behaviour all presuppose consciousness. (Elster 1983,
> 71–72)

Unfortunately for Elster, the trend in cognitive, evolutionary, and social psychology has been steadily away from this view toward ever more sophisticated, special purpose, unconscious information processing systems in both humans and other animals—so much that the concept of consciousness has become precarious and speculative (Allman 1999; Damasio 1994; Deacon 1998; Rolls 1998). Individual behavior is replete with latent functions, before we even proceed to the level of groups.

With respect to group psychology, fuzzy concepts of "social organisms" and "group minds" were easy prey for the seemingly hard-headed individualism and reductionism of the last few decades. However, I have already shown how the circuitry that we associate with "mind" can exist among a network of individuals as easily as among a network of neurons when groups are the units of selection. Examples are already known for the social insects, and finding comparable examples for humans might be primarily a matter of deciding that they are possible before we can see them clearly.[4] In any case, the more individuals act as participants in a group mental process, the less likely they are to be consciously aware of the process.

With respect to ongoing evolutionary processes, I have already shown in chapter 1 that what we loosely define as "culture" is in part a Darwin machine that produces phenotypic variation and heritability at the group level. If cultural phenotypic variation has functional consequences, it is hard to avoid the conclusion that at least some properties of present-day cultures owe their existence directly to the winnowing process of natural selection. In addition, it would be naive to expect people always to be aware of this source of function in their lives.

Since cultural adaptation as a direct product of cultural evolution has always been acceptable evidence for functionalism, it may surprise some readers that compelling examples can be found in the social sciences literature. Perhaps the best example is the Nuer, the African pastoralist tribe whose religion was described earlier in this chapter (summarized by Kelly 1985).[5] Linguistic evidence shows that the Nuer were historically derived from the Dinka, a neighboring tribe that shared the same environment and subsistence economy. Nevertheless, the Nuer and Dinka differed culturally in ways that gave the Nuer both a stronger incentive to raid cattle and the ability to field a larger fighting force in intertribal conflict. As a result, the Nuer were in the process of replacing the Dinka when they were contacted by anthropologists in the 1800s. If the "Nuer conquest" (actually the cumulative result of many independent raids and not an active campaign) had not been halted by the British and Egyptian administration, there would probably be no Dinka tribe for anthropologists to study.

The features of Nuer culture that accounted for their need to expand and their competitive superiority are complex and multifaceted, including brideprice customs that influenced herd management practices and a kinship system that enabled settlements to join together in war even after they had stopped interacting with respect to subsistence. Not all cultural differences between the two tribes are functional; there is plenty of room for what in evolutionary terms would be called byproduct explanations. Nevertheless, decades of anthropological research have established a convincing example of cultural replacement that can be causally attributed to cultural differences in which even the phylogenetic relationship of the two cultures is well understood. This example is generally thought by anthropologists to represent a common process of cultural replacement, even if other examples have not been documented to the same extent. As Darwin said, at all times throughout the world, tribes have supplanted other tribes. In addition, there is not the slightest hint that either the Nuer or the Dinka were aware of their differences or tried to consciously manage their cultures to succeed in inter-tribe competition. The cultural differences simply emerged and were stable enough to persist, with important functional consequences.

Closer to home are the cultural differences among regions of the United States that are derived from differences among the original colonists, and which remain discernible after centuries (Fischer 1989; Nisbett and Cohen 1996; Phillips 2000). It is likely that the outcome of the American Civil War was based as much on the long-term consequences of cultural variation as on environmental differences between the North and South. A third example is Putnam's (1993) exquisitely documented study of the patterns of cultural variation in Italy—patterns that have remained stable for a millennium, with profound functional consequences in the context of today's socioeconomic environment. Putnam's study illustrates the halting and inefficient nature of cultural evolution in addition to some of its basic ingredients. Cultural evolution may be fast in comparison to genetic evolution, but it can still require many human lifetimes. In addition, cultural variation does not always result in cultural evolution. Majority effects, spatial effects, and other factors complicate the concept of fitness for cultural and biological evolution alike, as I discussed in chapter 1 (see also Michod 1999a).

In none of these examples is there the slightest hint that people are aware of the functionally important features of their cultures, nor is there any compelling theoretical reason to expect them to. Indeed, Putnam (1993) pessimistically concludes that the deep structure of many contemporary human societies might prevent them from achieving the social goals for which they are intentionally striving. Similarly, Tocqueville (1835) observed that although the Mexican constitution was patterned after the American constitution, Mexico remained very different from the United States. Something mysterious that we don't understand, which in our ignorance we refer to as "custom" or "culture," is so important that it accounts for the fate of whole peoples and nations. It is a delusion that everything functional in human affairs is consciously intended. Evolutionary biologists and human social scientists should work together to understand the evolutionary processes that create latent functions in our species.

RATIONAL CHOICE THEORY REVISITED

Our progress so far can be summarized as follows: I began by describing rational choice theory as one of the most vigorous approaches to religion in the social sciences. It seems to reject functionalism in favor of a byproduct theory of religion, based on the general human tendency to seek explanations and employ cost-benefit reasoning. Then I examined the rejected tradition of functionalism, arguing that rumors of its death have been greatly exaggerated. The anthropologists who followed Durkheim did not reject functionalism on the basis of their greater field experience, and the most important objections raised by philosophers and social scientists must be revised on the basis of the evolutionary principles outlined in chapter 1.

Now I will show that the byproduct theory of religion is not nearly as robust as it is often taken to be. Before I criticize the rational choice literature, however, I must acknowledge some of its strengths. In may ways, it has been a productive research program and a model of scientific inquiry. It is a great achievement to develop a theory of religion based on the same principles that govern secular behavior. A good theory generates many testable hypotheses. A really good theory passes at least some of the tests. The rational choice theory of religion qualifies as a good theory by both of these criteria.

What, then, is left to criticize? There are a number of ways to bring religion under the umbrella of rational choice theory. One is to explain religion as the economic mind spinning its wheels to get what it can't have (a byproduct). Another is to explain religion as more genuinely utilitarian, producing resources that can be had but only through the beliefs and social organization provided by religion or a comparable system (an adaptation). Stark's formal theory of religion is strongly committed to the first approach, as his own list of propositions attests. However, the larger rational choice literature on religion, including some of Stark's own work, is not so committed and slips easily between these two forms of explanation.

Just as evolutionary theory can explain religion in many different ways, which I classified in table 1.1, many economic theories of religion exist that must be distinguished from each other. The individual who maximizes his own self-interest lies at the core of economic theory, which bears a superficial resemblance to individual selection theory in evolution. However, economics also offers a theory of the firm, in which groups are envisioned as the adaptive units. Mother Teresa poses no problem for economic theory because she employs cost-benefit reasoning to maximize her own peculiar utility of helping others (Kwilecki and Wilson 1998). If I sacrifice my time and hard-earned money praying to imaginary gods for resources that are not forthcoming, I fall within economic theory because I am employing cost-benefit reasoning, based on my beliefs (Stark 1999).

Clearly, economic theory shares with evolutionary theory the capacity to explain vastly different concepts of religion—but that is a weakness as much as a strength. If a theory can't distinguish among such different concepts and pit them against each other with empirical tests, what good is it? To make progress we must take a stand on a specific concept. I am taking a stand on the concept of religious groups as adaptive units, which allows me to aim a highly specific criticism at the rational choice literature: It *seems* to reject the concept I am evaluating, when in fact it does nothing of the sort. The situation is strikingly similar to the rejection of group selection in evolutionary biology in favor of theories that in fact included group selection within their own structure.

This problem pervades the rational choice literature but can be illustrated with a single example. In an article entitled "Voodoo Economics? Reviewing the Rational Choice Approach to Religion," Iannaccone (1995) compares the rational choice approach with alternative approaches as follows:

> The basic issue, after all, is whether people attend to costs and benefits and act so as to maximize their net benefits. The principal alternative is unreflective action based on habit, norms, emotion, neurosis, socialization, cultural constraints, or the like—action that is largely unresponsive to changes in perceived costs, benefits, or probabilities of success. (81–82)

> For decades, scholars have scrutinized religion from every angle except that of rational choice. Explanations of religious phenomena have stressed socialization, indoctrination, neurosis, cognitive dissonance, tradition, deviance, deprivation, functionalism, the role of emotions, the impact of culture and more. But rarely has anyone viewed religion as the product of cost-benefit decisions, and formal models of the religious behavior (rational or otherwise) scarcely exist. (86)

The appearance of functionalism so far down the list of alternatives shows how low its fortunes have sunk. In another passage Iannaccone includes "boundary maintenance" (82) as a specific function of religion that can be discarded along with everything else in this giant house-cleaning of the social sciences. Against this background, consider Iannaccone's own work. In two influential articles, he shows how seemingly inefficient and bizarre features of religion such as distinctive dress, dietary restrictions, and costly sacrifice can be adaptive (he uses the word "rational") by turning the religion into an exclusive club that excludes free-riders (Iannaccone 1992, 1994). Those who seek to selfishly exploit the church will be deterred by the costs and lost opportunities of membership. Those who do become members will work hard for the common good because their other options have been so circumscribed by their religion. They will benefit from their efforts because free-riders have been weeded out of the group. Iannaccone's theory accounts for the paradoxical fact, first noted by Kelly (1972), that churches are strong and vigorous in direct proportion to their strictness.

Iannaccone's theory is elegant, and I will discuss it in more detail in chapter 5. But is it a rational choice theory of religion that stands in contrast to an earlier, functionalist theory? For Iannaccone, the answer is obviously "yes" because cost-benefit reasoning is required for the free-riders and club members to respond to the religion as predicted by the theory. Fair enough. But how did the religion acquire its structure that adaptively constrains the choices of utility-maximizers in just the right way? We must explain the structure of the religion in addition to the behavior of individuals once the structure is in place. Were the bizarre customs consciously invented by rational actors attempting to maximize their utilities? If so, why did they have the utility of maximizing the common good of their church? Must we really attribute all adaptive features of a religion to a psychological process of cost-benefit reasoning? Isn't a process of blind variation and selective retention possible? After all, thousands of religions are born and die without notice because they never attract more than a few members (Stark and Bainbridge 1985). Perhaps the adaptive features of the few that survive are like random mutations rather than the product of rational choice.

Were it not for the vagaries of history and the shifting fashions of intellectual thought, Iannaccone's theory would qualify as a fine example of functionalism. Durkheim never denied cost-benefit religious reasoning. Recall the passage quoted earlier, which begins: "That man has an interest in knowing the world around him and that, consequently, his reflection was quickly applied to it, everyone will readily accept." He went further to say that religious belief is abandoned when it fails to have secular utility. It was Durkheim's insistence on secular utility that led him to reject the byproduct theories of his day and to explain religion as fundamentally functional for its members. The challenge for Durkheim was to interpret the many features of religion that appear to lack function as having latent functions. Iannaccone could accurately be described as a modern Durkheim who has revealed the latent functions of religion, including the all-important function of boundary marking, that serve to increase the collective utility of its members. Instead, he regards himself as a rational choice theorist and lumps boundary marking in particular and functionalism in

general along with a meaningless jumble of other concepts as outside rational choice theory. The fact that Iannaccone's theory explains how religion provides real benefits for its members (an adaptation) rather than
the illusion of getting what they can't have (a byproduct) is not perceived
as a problem. Who needs byproduct theories or functionalism when economics offers the theory of the firm?

THE FUTURE OF FUNCTIONALISM

I began this book by saying that many subjects have a long, hard road to
travel before they can advance by way of the textbook portrayal of the
scientific method. Functionalism is such a subject. Over a century of effort
hasn't even come close to resolving some of the most fundamental questions that can be asked about human behavior and social organization.
The worst that can be said about functionalism is that it failed to fulfill
itself as a research program during its heyday. Functionalism wasn't falsi-
fied; it merely went out of fashion.

Two major developments in intellectual thought may allow functionalism to succeed as a research program in the future, despite its past failures.
The first is progress in evolutionary biology, which provides the foundation for functional explanations of all kinds. When Evans-Pritchard (1965)
reviewed theories of primitive religion, he described Durkheim's theory as
"sociological" in contrast to a different and rejected class of "evolutionary"
theories. What Evans-Pritchard meant by evolution would be unrecogniz-
able today. Advances include not only multilevel selection theory (a very
recent development) but also the integration of ecology, evolution, and
behavior, the mature empirical study of adaptations, and modern evolutionary approaches to human behavior. If evolution is the foundation of
functionalism, then there is a new foundation upon which to build.

The second major development is the unification of the human social
sciences, which I described earlier as a vast archipelago of disciplines that
only partially communicate with each other. My status as an outsider
makes it easy for me to island-hop the social sciences, and I am struck by
the lack of consistency. To pick an example, anthropologists interested in
cultural change, even those who regard themselves as anti-Darwinian, tend
to emphasize non-intentional processes while neglecting the possibility
of rational choice. Cultures don't choose their destiny; it simply happens
to them through processes of which they are largely unaware. To question
this largely unstated assumption, Boehm (1996) searched the ethno-

graphic literature for case studies of how cultures respond to emergency events, such as warfare or natural disasters. In each case, the people responded by meeting as a group to discuss their options in a highly practical and rational fashion. Boehm's demonstration of a rational component to cultural change was new and important against the background of his particular discipline, although he would never claim that rational planning accounts for all cultural change. Contrast this configuration of ideas with Elster's extraordinary statement that "intentional explanation is the feature that distinguishes the social sciences from the natural sciences" (1983, 69). Then contrast Elster's reliance on conscious thought with the brain and behavioral sciences literature in which discussions of consciousness are reserved for the final chapters of books, with profuse apologies for their tentative nature (e.g., Rolls 1998)!

You know there is a problem when one man's heresy is another man's commonplace. It signals the need to step back and rebuild the social sciences from first principles, making the various subdisciplines consistent with each other and with evolutionary biology. One contribution that evolutionary biology can make to this enterprise is the classification outlined in table 1.1, which makes a fundamental distinction between functional vs. nonfunctional explanation and a secondary distinction between individual vs. group-level functional explanation. When functional explanations are warranted, the distinction between ultimate and proximate causation becomes critical. These are among the most basic distinctions that can be made from an evolutionary perspective, and they need to be preserved within the social sciences. As we have seen, rational choice theory slides so easily among these categories that they are scarcely recognized as categories. All subdisciplines of the social sciences need to appreciate that functional explanations must be handled with care, and that group-level functional explanations require the greatest care of all. Human groups cannot lightly be described as adaptive units, but if they can be rigorously shown to function as adaptive units, that will be a major scientific accomplishment.

A second contribution that evolutionary biology can make to rebuilding the social sciences is to provide a more sophisticated theory of psychology. As we have seen, rational choice theory is often presented as a psychological theory that attempts to explain the length and breadth of human nature with a few axioms about how people think. Alas, physics might be reducible to a few fundamental laws but not psychology. The human mind is a melange of adaptations and spandrels that have accumulated over millions of years, during which both culture and life in groups have been integral parts of the evolutionary process. Cost-benefit reasoning is an important part of human (and animal) psychology but not the only part. Tooby and Cosmides (1992) criticized learning theory in psychology for achieving the illusion of physics-like generality only by losing the ability to predict the specific content of what animals learn. The same criticism applies with equal force to rational choice theorists, who generally do not even attempt to predict the specific content of the utility that is supposedly maximized. This weakness, and evolutionary biology's strength in comparison, will become evident in subsequent chapters.

There was a time when individualism reigned supreme both in evolutionary biology and in the human social sciences, creating an image of the individual as the only adaptive unit (or rational actor) in nature and of the group as merely a byproduct of what individuals do to each other. Those days are over. I stated at the beginning of chapter 1 that science is not a frictionless pendulum destined to swing forever between extreme positions. Evolutionary biology is settling into a middle position that acknowledges the potential for adaptation and natural selection at all levels of the biological hierarchy, especially in the case of human evolution. Group-level adaptation is here to stay in evolutionary biology, and the human social sciences must follow suit to remain true to first principles. Now that the organismic concept of groups has emerged from its intellectual wilderness, let us see how well it describes the nature of religion.

CHAPTER 3

CALVINISM

An Argument from Design

> The major transitions in evolutionary units are from individual
> genes to networks of genes, from gene networks to bacteria-like
> cells, from bacteria-like cells to eukaryotic cells with organelles,
> from cells to multicellular organisms, and from solitary organisms
> to societies.
>
> —Michod 1999b, 60

Now at last we can proceed to our main goal of evaluating the organismic concept of religious groups as a serious scientific hypothesis. Thinking of a religious group as like an organism encourages us to look for adaptive complexity. Organisms do not maintain themselves and respond in just the right way to the challenges of their environment by chance. Mechanisms are required that are often awesome in their sophistication when sufficiently understood. Religions are certainly complex in their many beliefs and social practices, but are they adaptively complex? Can gods, rituals, and sacrifices actually be interpreted as a "social physiology" that enables a community unified by a religion to increase its collective secular utility, as Durkheim put it? If so, what are the processes that create adaptive complexity? Is it blind evolution, conscious intentional thought, cognitive processes that operate beneath conscious awareness, or all of the above?

These are risky questions whose answers can easily disprove our hypothesis. There are countless ways to be nonadaptively complex and only a few ways to be adaptively complex. Inside and outside the ivory tower, religion is often portrayed as costly for the believer, delivering at best only vague psychic benefits in return. Nonadaptation hypotheses are fully plausible, no matter whether presented in evolutionary terms, economic terms, or as a cocktail conversation. Thus, we have alternative hypotheses that make different predictions about observable aspects of religion. Which hypothesis is correct, or to what degree do they both contain a measure of truth?

Evolutionary biology offers a theoretical framework for thinking about these questions, but it also offers a set of empirical methods for completing the circle of scientific inquiry, from hypothesis formation to hypothesis testing and back again. I will attempt to study religious groups the way I and other evolutionary biologists routinely study guppies, trees, bacteria, and the rest of life on earth, with the intention of making progress that even a reasonable skeptic must acknowledge.

Science is often associated with complicated instruments and questions so specialized that a Ph.D. is required even to become interested. Evolutionary science often heads in this direction, but its core is a detailed understanding of organisms in relation to their environments. The foundation of this knowledge was provided by eighteenth- and nineteenth-century naturalists, most of whom believed that they were studying God's handiwork. Their careful observations and ingenious experiments without complicated instruments were sufficiently accurate to provide compelling evidence for Darwin's theory of evolution, even though the evidence was gathered with a completely different designing agent in mind. The habits of animals and their philosophical implications (science and philosophy had not yet differentiated) had a universal appeal that attracted a large audience without requiring specialized knowledge.

I mention all of this to suggest how much scientific progress can be made on the subject of religion without the trappings of modern science. I will survey the modern social scientific literature on religion in chapter 5, but we don't need to measure religiosity to the third decimal place or compare differences with multivariate statistics to make headway. We need to begin with a detailed understanding of religious communities in relation to their environments, and the foundation of this knowledge is provided by legions of scholars who have been studying religions with the same care and integrity as the early naturalists. Religious scholars are the natural historians for our subject. It doesn't matter that they have not been guided by an evolutionary perspective, any more than the early naturalists were guided by an evolutionary perspective. As long as the information is good, it can be used to test any hypothesis that calls upon it. Potentially, we can make as much progress on the strength of good old fashioned religious scholarship as Darwin made on the strength of good old fashioned natural history.

In this chapter, I will attempt to understand a single religious community in relation to its environment from an evolutionary perspective. Many examples would have sufficed, and others will be reviewed in the next chapter, but I chose John Calvin's brand of Christianity as it was instituted in the city of Geneva in the 1530s. My choice is based in part on the fact that Calvinism is a relatively new religious system and one sufficiently important to have attracted a great deal of scholarly attention. As I have stressed repeatedly, there is more to evolution than genetic evolution, and Calvinism provides an opportunity to study cultural adaptation to recent environments rather than genetic adaptations to ancient environments. As with all evolutionary analysis, however, we need to begin with a detailed picture of the organism in relation to its environment.

THE NATURAL HISTORY OF CALVINISM

John Calvin (1509–64) was a young French theologian and lawyer during the early days of the Reformation. He did not start out as a radical thinker. His early ambition was to succeed as a religious scholar in the newly formed tradition of humanism, which used the Greek and Latin classics as a source of inspiration rather than the more modern commentaries of the scholastic tradition (McGrath 1990).

The early Reformation was marked by many efforts to reform the Catholic Church from within, in addition to Luther's more daring break. At the time, it was difficult for individuals to predict whether their position would be regarded as an acceptable internal reform or a heresy worthy of banishment or death. The distinction was made as much on the basis of internal power struggles as on the basis of ideology. Calvin found himself on the losing end of such a struggle in 1533, when Nicolas Cop, the newly elected rector of the University of Paris, devoted his inaugural address to the need for reform and renewal within the church. It is possible that Cop's speech was actually written by Calvin, since one of the two existing copies is in his hand (McGrath 1990). In any case, even though Cop's ideas were mild in comparison with Luther's, his address provoked outrage. He was immediately replaced as rector and fled Paris to avoid arrest. Calvin also left Paris, probably wisely, since the authorities took action against numerous "Lutheran" sympathizers in the wake of the Cop affair, and Calvin almost certainly would have been among them had he remained.

Calvin still was not reform-minded when he found refuge (along with Cop) in the Swiss city of Basel in 1535. There he observed the dramatic events taking place elsewhere, including the execution of his friend Etienne de la Forge and the description of the French reformers as Anabaptists. This charge was extremely serious because the Anabaptists were far more radical than either Luther or the French reformers. To be labeled an Anabaptist was to be accused of treason, punishable by death. The charge was also outrageous because the French reform movement theologically bore little relationship to the Anabaptist movement. Calvin himself had written a treatise against the Anabaptists. Stung as much by the poor scholarship of the accusation as by its implications, Calvin wrote a book describing the foundations of his religious faith that became his greatest work, *The Institutes of the Christian Religion.*

Calvin still was not reform-minded when he passed through Geneva en route to Strasbourg in 1536. The city of Geneva had recently expelled the Roman Catholic Church as part of becoming independent from the duchy of Savoy. It craved independence but was totally dependent for military support upon the Swiss Confederacy, especially the city of Berne. The Swiss reform movement spread to Geneva but lacked organization. In addition, the city was governed by a democratically elected council that had only recently gained independence from the Catholic Church and was not about to yield to a new religious authority.

Calvin was converted from a bookish scholar to a religious activist by Geneva's two leading reformers, Guillaume Farel and Pierre Viret. I cannot improve on Calvin's own description of the event (quoted in McGrath 1990, 95):

A little while previously, popery had been driven out by the good man I have mentioned [Farel] and by Pierre Viret. Things, however, were still far from settled, and there were divisions and serious and dangerous factions among the inhabitants of the town. Then someone, who has now wickedly rebelled and returned to the papists, discovered me and made it known to others. Upon this Farel (who burned with a marvelous zeal to advance the gospel) went out of his way to keep me. And after having heard that I had several private studies for which I wished to keep myself free, and finding that he got nowhere with his requests, he gave vent to an imprecation, that it might please God to curse my leisure and the peace for study that I was looking for, if I went away and refused to give them support and help in a situation of such great need. These words so shocked and moved me, that I gave up the journey I had intended to make. However, conscious of my shame and timidity, I did not want myself to be obliged to carry out any particular duties.

As McGrath (1990, 96) has noted, "Precisely what Farel saw in him we shall never know." However, Calvin proved his worth only a few weeks later in a public disputation that was to decide whether the city of Lausanne would remain Catholic or become reformed. The Catholics accused the reformers of impiety for ignoring the Christian writers of the first five centuries. The scholarly Calvin replied that not only did the reformers appreciate them, but they also knew more about them than the Catholics. He then proved his point by recounting so many facts from memory that the opposition was humbled. Calvin's scholarship and formidable writing skills became powerful tools in his new role as social reformer.

Calvin is sometimes portrayed as a religious tyrant who ruled Geneva with an iron fist. Nothing could be further from the truth, as virtually all Calvin scholars appreciate. Calvin had no formal authority as far as the civil government was concerned and never even became a citizen of Geneva. His plan for reform was initially rejected, and both Calvin and Farel were expelled from Geneva in 1538, only to be invited back three years later. The city council continued to oppose Calvin on many issues until 1555, when wealthy refugees, most of whom supported Calvin, were allowed to purchase the status of bourgeois that enabled them to vote. Thereafter, the relationship between the reformed church and the city council became more symbiotic, but nevertheless Calvin continued to function in a purely advisory role. In addition, Calvin did not even rule his own church with an iron fist. He shared the status of pastor with several others who made decisions on a consensus basis. The decision-making structure of Calvin's church will be discussed in more detail below.

Calvin's influence on Geneva cannot be attributed to his personal power or charisma, although the moral example that he set for others may have been a factor. Instead, we must look to the belief system and the social organization that he established, which caused a city of roughly 13,000 souls to function more effectively than it did before. Indeed, reform-minded people from all over Europe flocked to Geneva to learn and export the secrets of its success. The fact that the exported versions of Calvinism also tended to succeed points to the belief system and social organization, rather than the person, as the cause of the success.

By Calvin's own account, his primary challenge was to unite the fractious city of Geneva into an effective corporate unit. This challenge placed the church in an uneasy symbiotic relationship with the Genevan civil government, a relationship that wavered between cooperation and conflict with the church the subordinate member of the pair. The city itself was the subordinate member of an uneasy alliance with one larger political power (the

Swiss Confederacy) against another (the duchy of Savoy). This is the complex social and political environment inhabited by Calvin's church.

AN ADAPTATIONIST INTERPRETATION OF CALVINISM

Studying adaptations at the appropriate scale

Our next step is to see if the detailed properties of Calvin's church can be interpreted as adaptations to its environment. First, however, some general issues about spatial and temporal scale need to be discussed. Calvinism is not the product of a single man, nor is it something that remained constant through time. Calvin was obviously building on a huge foundation of Judaeo-Christian belief, and what he implemented in Geneva was influenced by other people and institutions, such as Geneva's city government. Calvinism also changed after Calvin's death. According to McGrath (1990, 209–11), the emphasis on predestination that is often associated with Calvinism did not become prominent until after Calvin's death and served a specific purpose: to demarcate so-called "Calvinists" from rival Protestant groups (note the functional nature of this explanation). When we study the adaptedness of Calvinism to its environment, should we be looking at the coarsest spatial/temporal scale of the Christian tradition, a somewhat finer scale of the Calvinist tradition within Christianity, or the even finer scale of the church of Geneva during Calvin's lifetime?

The same questions confront evolutionary biologists studying nonhuman species. When we study the adaptedness of guppies to their environment, should we be looking at the coarse scale of the larger taxonomic units to which guppies belong (genus *Poecilia*, family Poecillidae, order Cyprinidontiformes), the somewhat finer scale of guppies as a species, the even finer scale of guppies in a single stream, the yet finer scale of upstream vs. downstream populations, or perhaps the ultimate fine scale of microhabitats within a single pool? These are empirical questions that must be answered by painstaking research, and the answers can differ on a trait-by-trait basis. The trait of giving birth to live young is characteristic of the entire family that includes guppies. It is probably an adaptation, but not one that responds to local selection pressures. The trait of offspring size at birth is more variable within guppy populations and does respond to local selection pressures; downstream guppies are smaller at birth (and there are more of them) than upstream guppies, which makes perfect sense as adaptations to the presence and absence of predators.

Looking for adaptation at too fine a scale leads to a simple falsification of the hypothesis. For example, suppose I predict that guppies close to the edge of a single pool (a low-predation microhabitat) give birth to larger young than guppies in the middle of the same pool (a high-predation microhabitat). If I fail to find the predicted difference, it is probably because adaptation and natural selection have not occurred at such a fine scale. Looking for adaptations at too coarse a scale leads to a different kind of error; a failure to see adaptations that do, in fact, exist. Suppose I think that guppies are well adapted as a species but I do not appreciate the local adaptations that exist within single streams. In that case, I will not distinguish between upstream and downstream populations, and the adaptive variation that exists within a single stream—males that vary in their color, babies that vary in their birth weight—will appear as senseless noise. Adaptations at a given spatial and temporal scale become invisible when studied at a larger spatial and temporal scale.

The historical trend in evolutionary biology has been in the direction of discovering the power of adaptation and natural selection at increasingly fine spatial and temporal scales. During the 1940s the differences between the various species and subspecies of finches on the Galapagos Islands were explained as a product of genetic drift. The British ecologist David Lack (1961) became famous for explaining these differences in terms of adaptation and natural selection, but even he would be astonished to discover the even finer spatial and temporal scale of natural selection that is recounted so well in Jonathan Weiner's *The Beak of the Finch* (1994).

Thinking of religious groups as adaptive units comparable to guppies and finches is so new that it is impossible at present to identify the appropriate spatial and temporal scale. However, one reason that religious belief often appears senseless to outsiders may be because it is being approached at too coarse a scale (e.g., the concept of Christian forgiveness, discussed in chapter 6). Adaptive pattern may emerge with more fine-grained analysis. I will therefore begin at a small scale by evaluating the adaptedness of a given form of the Christian religion for the particular community of people practicing that form. In the present case, I will attempt to evaluate the adaptedness of Calvinism for the inhabitants of Geneva in the mid-1500s. My analysis will include features of Calvinism that are shared by other denominations (analogous to live birth in guppies) as well as features that distinguish Calvinism from other forms of Christianity.

Catechisms as data

Examining a religion for its functional properties might seem like a daunting task. However, many modern forms of Christianity are summarized in catechisms, which are explicitly taught to those entering the faith. One of Calvin's first acts was to write a catechism and to insist that it be learned by every inhabitant of Geneva. Calvin scholars agree that his catechism provides an excellent summary of the views developed at much greater length in the *Institutes* and his many other works. In addition, it is the most relevant summary for our purposes because it is what the inhabitants of Geneva actually learned, and therefore it presumably had the greatest impact on their behavior.[1]

Catechisms are a potential gold mine of information for the evolutionary study of religion. They may truly qualify as "cultural genomes," containing in easily replicated form the information required to develop an adaptive community. They are short enough for detailed analysis, and many religious denominations have them, enabling the comparative study of religious organizations. Finally, single denominations periodically revise their catechisms, providing a neatly packaged "fossil record" of their evolutionary change. It would be hard to imagine a better historical data base.

I will now attempt to analyze the first catechism that Calvin wrote for the city of Geneva in 1538, based on F. L. Battle's translation (reprinted in Hesselink 1997; all page numbers refer to this edition). My objectives can be stated in the form of two simple questions: How would a person who learned and believed Calvin's catechism be motivated to behave? What are the specific design features of the catechism that accomplish its effect on behavior?

Calvin's catechism

Calvin's catechism places equal emphasis on peoples' relationship with God and their relationship with other people. Calvin himself makes this distinction:

> In God's law is given the most perfect rule of all righteousness, which is for the best of reasons to be called the Lord's everlasting will. For in its two tables has been included fully and clearly all that we need. The first table has in a few commandments set forth the worship appropriate to his majesty; the second, the duties of charity owed to one's neighbor. (11)

From the functional standpoint, both the concept of God and his relationship with people need to be explained as an adaptation for regulating human conduct. Thus, the entire catechism must be understood in terms of people-people relationships, at least as a working hypothesis. However, it is convenient to retain the God-people and people-people distinction for purposes of analysis. Describing how God expects people to behave toward each other is relatively simple because it is explicitly stated in the catechism. Why the concept of God and a particular God-people relationship is needed to encourage the behaviors is a deeper, although still answerable, question.

The people-people relationship

A list of behaviors prescribed by Calvin's catechism is shown in table 3.1. It includes the familiar injunctions of the Ten Commandments, which, like live birth in guppies, are not expected to vary within the Judaeo-Christian tradition. It also includes the organism metaphor that forms the inspiration for this book. Finally, the list includes behaviors that appear tailored to the local social and political environment. Taking these in reverse order, Calvinism instructs the believer to be a solid citizen and not to rebel against political authority, as described in the following passage:

Table 3.1 Elements of the people-people relationship specified by Calvin's catechism of 1538

1. Obey parents
2. Obey magistrates
3. Obey pastors
4. Abandon self-will
5. Do unto others as you would have them do unto you
6. Behave as a single organism
7. Don't harm or kill your neighbor
8. No lewdness and sex only in marriage
9. No theft, either by violence or cunning
10. Don't swear false oaths
11. Don't bear false testimony
12. Pay taxes and perform other civic duties
13. Behave in a civil manner

> Not only should we behave obediently toward those leaders who perform their office uprightly and faithfully as they ought, but also it is fitting to endure those who insolently abuse their power, until freed from their yoke by a lawful order. For as a good prince is proof of divine benefi- cence for the preservation of human welfare, so a bad and wicked ruler is his whip to chastise the peoples' transgressions. (38)

Magistrates are expected to be righteous, but a meek attitude is adopted toward those who are not. In fact, a wicked ruler is actually performing God's will by punishing members of the church for their own transgres- sions! This seems to provide a clear case in which a statement about God is used as a mechanism to invoke a behavior (meekness toward political authority) that is adaptive in the church's real-world environment.

Although Calvin's church had little control over political authority, it potentially had more control over its own leaders. If religion is primarily exploitative, allowing high-status members to gain at the expense of low- status members, we might expect the catechism to include a "loophole" for wicked pastors similar to that for wicked princes, but such is not the case:

> Therefore, pastors may dare boldly to do all things by God's Word, whose stewards they have been appointed to be. They may compel all worldly power, glory, wisdom, and loftiness to fall down and obey his majesty. Through it they are to command all from the highest even to the last. They are to build up Christ's household and cast down Satan's kingdom; to feed the sheep and slay the wolves; to teach and exhort the teachable; they are to accuse, rebuke, and subdue the rebellious and stubborn. But let them turn aside from this to their own dreams and figments of their own brains, then they are no longer to be considered to be pastors but rather as pestilential wolves to be driven out. For Christ does not com- mand others to be heard than those who teach us what they have taken from his Word. (36)

This passage provides our first hint that Calvin's church was designed to control the conduct of the shepherd as effectively as the conduct of the flock.

The organism metaphor encourages unity of purpose but also a divi- sion of labor to achieve shared goals that goes beyond the phrase "do unto others":

The whole number of the elect are joined by the bond of faith into one church and society and one people of God, which Christ our Lord is Leader and Prince, and, so to speak, Head of the one body, just as in him before the foundation of the world they were all chosen to be gathered into God's kingdom. Now this society is catholic, that is, universal, because there could not be two or three churches, but all God's elect are so united and conjoined together in Christ that as they are dependent upon one Head, they also grow together into one body, being joined and knit together as are the limbs of one body. They are made truly one since they live together in one faith, hope, and love, and in the same Spirit of God, called to the inheritance of eternal life. . . . Just as the members of one body share among themselves by some sort of community, each nonetheless has his special gift and distinct ministry. (25–26)

One of the hallmarks of Calvinism, well-recognized by Calvin scholars, is that it sanctifies the mundane occupations of life. A baker or farmer can feel an element of holiness similar to that of a priest because all are ministers, performing organ-like functions to sustain the body of the church.

Although the familiar injunctions of the Ten Commandments and the Golden Rule may appear too obvious to deserve comment, we should not be misled by their familiarity. The hypothesis I am seeking to test proposes that religion causes human groups to function as adaptive units. This hypothesis has many rivals, including religion as a tool of exploitation, as a cultural parasite, as a byproduct of cost-benefit reasoning—the list goes on and on. As we saw in chapters 1 and 2, intellectual trends in biology and the social sciences have made the organismic concept of religion an underdog hypothesis, even a heresy. The Ten Commandments and the Golden Rule may be familiar, but they are far more supportive of the organismic hypothesis than its rivals. The evolutionary study of human behavior includes other insights that are obvious in retrospect. From Daly and Wilson (1988) we learn that men are more violent than women and that parental abuse falls more heavily on step-offspring than on biological offspring. From Buss (1994) we learn that men especially value youth and beauty in women, while women especially value wealth and status in men. These results are important, despite their familiarity, because they emerge from a formal theoretical framework rather than from folk psychology and also because they are not explained and often are outright denied by rival perspectives in the social sciences. In our case, we are predicting that a religion instructs believers to behave for the benefit of their group, which is supported by the Ten Commandments and the Golden Rule, however familiar. It is better to be obvious than to be wrong.

In addition, these familiar injunctions include some features that warrant discussion from a group-level functional perspective. Communication is often assumed to be a cooperative activity that benefits everyone, senders and receivers alike. Evolutionary biologists have pointed out that honest, mutually beneficial communication cannot be axiomatically assumed (Dawkins and Krebs 1978). Language can be used for deception and exploitation as well as for cooperation. Honest communication, like any other form of prosocial behavior, is vulnerable to the fundamental problem of social life and requires special conditions to evolve. Calvin's catechism (and the entire Judaeo-Christian tradition) appears to recognize this by making false oaths and false testimony as sinful as murder and adultery.

The familiar Christian emphasis on forgiveness also deserves a closer look. "Turn the other cheek" is often interpreted as an invitation for exploitation, but at least in the case of Calvinism nothing could be further from the truth. In the first place, Calvin's catechism discusses forgiveness far more in the context of the God-people relationship (described below) than of the people-people relationship. In the second place, it provides a detailed procedure for punishing transgressions in which forgiveness is highly conditional upon repentance (also described below). What then, is the meaning of "Forgive us our debts, as we forgive our debtors":

> We petition that this pardon be granted "as we forgive our debtors," namely, as we spare and pardon those who have in any way injured us, either treating us unjustly in deed or insulting us in word. This condition is not added as if by our forgiveness which we bestow on others we merit God's forgiveness. Rather, this is set forth as a sign to assure us he has granted forgiveness of sins to us just as surely as we are aware of having forgiven others, provided our hearts have been emptied and purged of all hatred, envy, and vengeance. Conversely it is by this mark that those who are eager for revenge and slow to forgive and practice persistent enmity are expunged from the number of God's children, in order that they may not dare call upon God as Father and thus avert from themselves the displeasure they foment against others. (32)

Note that Calvin explicitly rejects the notion of bargaining with God, in contrast to Stark's conception of religious belief. My functional interpretation of this passage is that the church has taken over the role of maintaining social order from its members. In the absence of a strong church or comparable social organization, individuals must maintain their own social order, which leads to a limited amount of cooperation at a small scale but also to feuds and rivalries that are dysfunctional at a larger scale. The factions that plagued Geneva prior to Calvinism were cooperating internally but not with each other. A church that attempts to build a unified society at a large scale must suppress the self-help mechanisms that come all too naturally to its members. Members must forgive each other's trespasses, which will be punished at a higher level. As we shall see, this 'higher level' is not an imaginary hell, but a well-oiled mechanism that far surpasses self-help for punishing transgressions in the real world.

To summarize, Calvin's catechism provides a blueprint for human conduct that makes sense from a group-level functional perspective. The blueprint goes beyond general prescriptions to "do unto others" and is fine-tuned to its local environment. For any unit (individual or group) to function adaptively, its behavior must be highly context-sensitive. A blueprint for adaptive behavior must go beyond "do x" to "do x in this situation, do y in this other situation . . ." and so on. Calvin's catechism is impressively context-sensitive for a short document, and much more detail is provided in his other writings, as we shall see.

The concept of local adaptation allows religion to be studied from an evolutionary perspective using the same methods employed on nonhuman species, as I discussed in chapter 2. If the environment changes over time and space, and if religions adapt human groups to their environments, we should be able to predict the properties of religion at a fine spatial and temporal scale as surely as we can predict the properties of upstream and downstream guppies. I am restricting myself to a single religious population in this chapter, but I will attempt a more comparative analysis for the concept of forgiveness in chapter 6.[2]

The God-people relationship

Ask a person to do something and the most likely response will be "Why?" An adaptive belief system cannot simply provide a list of behaviors but must also justify them. It might seem that the justification could be factual and straightforward: "Do this because it is good for you." However, this approach is unlikely to succeed by itself for a number of reasons. First, it works best when the consequences of the behavior are well known: "Eat your spinach because it is high in iron and will make you healthy."[3] Often the consequences of behaviors are not well known, and the most obvious short-term consequences (the bitter taste of spinach) can lead to a different conclusion than the more subtle long-term consequences (the health effects). An adaptive belief system must cope with ignorance in its justification of behaviors.

Second, a belief system that is adaptive at the group level must cope with the problem of cheating, which benefits some individuals at the expense of others within the group. Cheating is genuinely beneficial for the cheater (when he or she gets away with it), and therefore cannot be argued against on the basis of personal benefit. The same point can be made in terms of the "veil of ignorance" that Rawls (1971) used to explain the concept of justice. Ask self-interested people to design a society, subject to the constraint that they will be placed at random within the society, and they will design a just society. However, once placed within the society, they are subject to a different set of constraints and may well want to destroy what they previously created. This problem, which lies at the heart of multilevel selection theory, makes it difficult to justify the behaviors that constitute an adaptive group in terms of personal benefit.

Third, an adaptive belief system must be economical. The beliefs that justify the behaviors must be easily learned and employed in the real world. A fictional belief system that is user-friendly and that motivates an adaptive suite of behaviors will surpass a realistic belief system that requires a Ph.D. to understand and that leads to a paralysis of indecision.

Fourth, a fictional belief system can be more motivating than a realistic belief system. Imagine two individuals competing for a common resource. Even though the facts of this situation are easy to comprehend, regarding one's enemy as inhuman can be more motivating than regarding one's enemy as just like oneself.

Fifth, a fictional belief system can perform the same functions as externally imposed rewards and punishments, often at a much lower cost. For example, the usual means of raising money to serve the common good is in the form of taxes. Unfortunately, individuals who avoid paying taxes without punishment are always better off in material terms than solid citizens within the same group. Cheating can be prevented by punishment, but implementing a system for detecting and punishing cheaters can itself be costly. Another solution is to manipulate the cost of cheating in the mind of the average citizen. Groups governed by belief systems that internalize social control can be much more successful than groups that must rely on external forms of social control.

For all of these (and probably other) reasons, we can expect many belief systems to be massively fictional in their portrayal of the world (Wilson 1990, 1995). As I discussed in chapter 1, their adaptedness must be judged by the behaviors they motivate, not by their factual correspondence to reality. Returning to Calvin's catechism, our challenge is to interpret the concept of God and his relationship with people as an elaborate belief system designed to motivate the behaviors listed in table 3.1.

It is common for nonreligious folk to regard heaven and hell as gimmicks for promoting good conduct. Evolutionary biologists also have speculated that the religious use of terms such as "father" and "brother" triggers kin-selected helping behaviors toward nonrelatives (Alexander 1987).[4] These suggestions are in the right direction but they do not go far enough. Taking the organism metaphor seriously may allow us to discover a more sophisticated "motivational physiology" than we have heretofore recognized.

Major elements of the God-person relationship in Calvin's catechism are listed in table 3.2. God is portrayed as an all-powerful being who created man in his own image. Instead of worshiping God with fitting gratitude, man arrogantly placed himself above God and had to be deprived of all glory to recognize its true source. All people share this original sin and are born thoroughly depraved. By themselves they have no free will to choose between good and evil; they are bent on evil. Their only path to redemption is to become convinced of their corrupt nature and the certain horrible fate that awaits them if they do not earnestly seek God. Then, however, they have opened "the first door into his Kingdom" by overthrowing "the two most harmful plagues of all, carefree disregard of his vengeance and false confidence in our own capacity."

Now God reveals his merciful side by forgiving the sins of those who approach him in the right way. The forgiveness is in no way deserved, for only Christ is pure enough to be without sin, but is due entirely to God's "inexpressible kindness." After all, he did not deprive us of glory to permanently forsake us but to show us the way back to him. The prevailing emotion of the true believer therefore is not fear of God's vengeance but joy at having been saved. Of course, joy and fear are inextricably linked. One is joyful to have been spared a horrible fate. One is grateful to the merciful God who grants what we do not deserve, but fearful of the same God who punishes those who do not enter the faith, or, even worse, enter and then leave it. God's curse is so great in this case that it extends to the fourth generation (great-grandoffspring), although his kindness is much greater and extends a thousand generations for those who keep the faith. God is described not only as a father who has our best interest at heart but also as a powerful and just Lord who is quick to punish those who

Table 3.2 Elements of the God-people relationship specified by Calvin's catechism of 1538

1.	Forgiveness of sins
2.	Faith
3.	Internalization
4.	Preparing for the second coming

do not earnestly try to seek him. To feel secure within this belief system, one must strive for perfection with all one's heart and soul, even though it is impossible to achieve.

A large portion of Calvin's catechism is devoted to establishing this system of beliefs. It should be obvious that anyone who takes it seriously will be placed in an exceptionally motivated and compliant frame of mind. Calvinism uses elements of human psychology that exist apart from religion as building blocks to build a new structure. For example, repentance and forgiveness are so basic to social relationships that they emerge from simple game theory models, as I will describe in chapter 6. They are surely part of the innate psychology that I discussed in chapter 1. Most people who violate a social norm feel apologetic and eager to make amends, while most people whose norms have been violated are willing to forgive after an appropriate period of repentance. However, repentance is only one of many states of mind that are evoked in response to particular situations. Most of the time we are unapologetic and cling tenaciously to our own agendas. Calvin's catechism is designed to make this one state of mind the normal human condition. Forgiveness of sins is a far more sophisticated piece of "motivational physiology" than the use of kinship terms or the simple fear of hell.

Faith is another facet of normal human psychology that is vastly expanded and modified by Calvin's catechism. One dictionary definition of faith is "confidence, reliance, belief, especially without evidence or proof." Faith is required for action in an uncertain world, but usually it is designed to be modified by experience. Faith in a spouse's fidelity is shaken and withdrawn by solid proof of an affair. Faith that a given part of the forest is teeming with game is gradually diminished by lack of hunting success. Calvin's catechism turns faith from a belief designed to be modified by experience into a fortress designed to protect the belief system from experience. The way that Calvin does this is actually very compelling. Remember that the human intellect is infinitely puny compared to God's. If we have difficulty discerning each other's will, what is our hope of discerning God's will? The very idea is part of the hubris that must be abandoned to approach God. It follows that all of life's afflictions have a purpose in God's plan, however incomprehensible to us. Our role is to be utterly confident in God's wisdom and to accept whatever he places upon us:

> If faith . . . is a sure persuasion of the truth of God, a persuasion that cannot lie to us, deceive us, or vex us, then those who have grasped this assurance expect that it will straightway come to pass that God will fulfill his promises, since according to their opinion they cannot but be true. (31)

> And whatever God does, let all his works appear glorious, as they are. If he punishes, let him be proclaimed righteous; if he pardons, merciful; if he carries out what he has promised, truthful. In short, let there be nothing at all wherein his graven glory does not shine, and thus let praises of him resound in all hearts and on all tongues. (30)

Skeptics often consider the unshakable religious faith expressed in these passages to be irrational, especially compared to the scientific method, which attempts to hold all articles of faith to the fire of evidence. However, if religious belief systems are to be studied from an evolutionary perspective, they must be evaluated by the appropriate criteria. If religious faith plays a role in motivating the behaviors in table 3.1, and if these behaviors cause the group to function as an adaptive unit, then faith counts as an adaptation. Faith that ignores experience can even count as rational, as economists use the word. Suppose that behavior 1 is adaptive in situation A but has no effect on fitness in situations B through Z. Why should the afflictions caused by B–Z lead to the abandonment of behavior 1? If the behaviors in table 3.1 benefit the group in many respects, why should they be abandoned because of plagues, droughts, invading armies, and other afflictions beyond the group's control? The ability of a belief system to survive these shocks is to be admired from an adaptationist perspective, not ridiculed. Furthermore, religious faith does not necessarily imply that religious groups cannot change their behavior on the basis of experience. Part of taking the organismic concept seriously involves thinking in terms of a "head" and a "body." The head of the church of Geneva was not Christ but a small group of pastors who met frequently to discuss alternative courses of action within the framework of their beliefs, resulting in policies that the body of the church was expected to accept on faith.[5] The anatomy of the church's head and the quality of its decisions will be described in more detail below.

Another pillar of the God-people relationship in Calvin's catechism is the internalization of the belief system. The ideal is a complete replacement of self-will with God's will, which becomes second nature. Calvin's explication of the tenth commandment makes this clear:

> By this commandment the Lord imposes a bridle upon all our desires which outstrip the bounds of charity. For all that the other commandments forbid us to commit in deed against the rule of love, this one prohibits from being conceived in the heart. Accordingly, by this commandment hatred, envy, and ill-will are condemned, just as much as murder previously was condemned. Lust and inner filth of heart, just as much as fornication, are forbidden. Where previous rapacity and cunning were restrained, here avarice is; where cursing, here spite is curbed. We see how general the scope of this commandment is and how far and wide it extends. For God requires a wonderful affection and a love of the brethren of surpassing ardor, which he does not wish even by any desire to be aroused against a neighbor's possessions and advantage. This then is the sum of the commandment: we ought to be so minded as not to be tickled by any longing contrary to the law of love, and be prepared utterly freely to render to each what is his. We must reckon as belonging to each man what we owe him out of our duty. (15)

Despite his own intellectual talents, Calvin scorned "bare knowledge of God or understanding of Scripture which rattles around the brain and affects the heart not at all" (18). Similarly, exemplary behavior by itself is nothing without an underlying "purity of conscience" (20). Of course, purity of conscience cannot fail to produce exemplary behavior, so the emphasis on faith rather than works is not a license for selfishness.

Internalization is encouraged by prayer, which is described in detail as a private relationship between each individual and God. A man of true faith sees himself as so empty that he must consult God for guidance, and God is so compassionate that he wants to hear from each and every one of us. Neglecting prayer would be like a person who neglects "a treasure, buried and hidden in the earth, after it had been pointed out to him" (28). We do not pray to boast about our accomplishments or to seek personal gain, but to express our innermost feelings:

Since prayer is a sort of agreement between us and God whereby we pour out before him all the desires, joys, sighs, and finally, thoughts of our hearts, we must diligently see to it, as often as we call upon the Lord, that we descend into the innermost recesses of our hearts and from that place, not from the throat and tongue, call God. Although the tongue sometimes contributes something in prayer, either to keep the mind intent by its exercise upon thinking about God, or to occupy that part of us expressly destined to proclaim God's glory, together with the heart, with meditating on the Lord's goodness—the Lord through his prophet has declared what value this has when mindlessly expressed, when he lays the gravest vengeance upon all who, estranged from him in heart, honor him with their lips. (28)

The Lord's prayer is provided as a model, a way to pray to God in God's own words, and indeed it serves as an admirable one-minute summary of the entire belief system. In his explication of the phrase "Lead us not into temptation," Calvin recognizes that the ideal of internalization is as unattainable as the ideals of behavior, and he even describes temptation as a kind of exercise that keeps us spiritually strong:

By this petition we do not entreat that we feel no temptation at all, for it is in our interest rather to be aroused, pricked, and urged by them more lest, with too much inactivity, we grow sluggish—just as the Lord daily tests his elect, chastising them by disgrace, poverty, tribulation, and other sorts of affliction. Rather, this is our plea: that along with the temptation he may bring it about that we may not be vanquished or overwhelmed by any temptations, but strong and vigorous by his power we may stand fast against all powers that attack us. Secondly, that received into his care and safekeeping, and sanctified by his spiritual gifts, fortified by his protection, we may stand unconquered by the devil, death, the gates of hell, and the devil's whole kingdom. We must moreover note how the Lord wills our prayers to be shaped by the rule of love, for he has determined that we are not to seek for ourselves what is expedient for us apart from concern for our brothers; but he bids us to be concerned about their good just as much as our own. (32)

As if these features of the God-person relationship were not enough to motivate behavior, Calvin believed (along with many others of his time) that the final struggle was at hand between God and the Antichrist, whom Calvin firmly believed was the Pope. However, Christ wanted his church to be rebuilt and restored before his return. This imparted not only a sense of urgency but also a sense that the past and present did not predict what is possible in the future. According to Wallace (1988, 40),

> In his *Institutes* the first of many attacks he launches against our human nature is against this tendency within us all to accept everything around us as being final and settled—to 'confine our minds within the limits of human corruption,' to be content with our 'empty image of righteousness' instead of seeking the reality, to become delighted merely when things around us are not so bad as they could become. He expected progress. He looked for results that would confound all human expectations, deserving the comment: "This is the Lord's doing and it is marvelous in our eyes."

To summarize, the God-people relationship can be interpreted as a belief system that is designed to motivate the behaviors listed in table 3.1. Those who regard religious belief as senseless superstition may need to revise their own beliefs. Those who regard supernatural agents as imaginary providers of imaginary services may have underestimated the functionality of the God-person relationship in generating real services that can be achieved only by communal effort. Those who already think about religion in functional terms may be on the right track, but they may have underestimated the sophistication of the "motivational physiology" that goes far beyond the use of kinship terms and fear of hell. Indeed, it is hard for me to imagine a belief system better designed to motivate group-adaptive behavior for those who accept it as true. When it comes to turning a group into a societal organism, scarcely a word of Calvin's catechism is out of place.

Social organization

No matter how powerful, a belief system by itself is probably insufficient to turn a group into a societal organism. A social organization is also required to enforce the norms and to coordinate the activities of those who abide by the norms. One of the most important conclusions to emerge from the major-transitions-of-life paradigm is that all adaptive units, including individual organisms, require mechanisms to prevent subversion from within. The same sentiment was expressed for religious groups by Calvin's contemporary Martin Bucer (1524; quoted in McNeill 1954, 80–81), who stated: "Where there is no discipline and excommunication there is no Christian community."

Table 3.3 Elements of the social organization specified by Calvin's catechism of
1538 and Ecclesiastical Ordinances of 1541

---

1. Policing the head of the church
2. Policing the body of the church
3. Decision-making
4. Coordination of community activities (health, education, welfare)

Calvin's catechism was accompanied by a second short document called the Ecclesiastical Ordinances that specified the social organization for the church of Geneva (included in Reid 1954; all subsequent page number refer to this edition). Four orders of office were specified: the pastors, the doctors, the elders, and the deacons. The duties of the pastors were "to proclaim the Word of God, to instruct, admonish, exhort and censure, both in public and private, to administer the sacraments and to enjoin brotherly corrections along with the elders and colleagues" (58). Pastors were required to have both a thorough knowledge of church doctrine and an excellent reputation. A candidate was first chosen by the existing pastors, was then presented to the city council for acceptance, and finally was presented to the congregation to which he would preach "in order that he be received by the common consent of the company of the faithful." This is another example of how Calvinism was structured to control the leadership as effectively as the average member of the church.

Pastors were not only selected with care but also were carefully supervised, with weekly meetings to insure purity of doctrine and quarterly meetings whose express purpose was for the pastors to criticize each other. Recall from chapter 2 that the Nuer also had ceremonies for the purpose of mutual criticism. Calvin was very specific about the appropriate conduct of the pastors:

> But first it should be noted that there are crimes which are quite intolerable in a minister, and there are faults which may on the other hand be endured while direct fraternal admonitions are offered.
>
> Of the first sort are: heresy, schism, rebellion against ecclesiastical order, blasphemy open and meriting civil punishment, simony and all corruption in presentations, intrigue to occupy another's place, leaving one's Church without lawful leave or just calling, duplicity, perjury, lewdness, larceny, drunkenness, assault meriting punishment by law, usury, games forbidden by the law and scandalous, dances and similar dissoluteness,

crimes carrying with them loss of civil rights, crime giving rise to another separation from the Church.

Of the second sort are: strange methods of treating Scripture which turn to scandal, curiosity in investigating idle questions, advancing some doctrine or kind of practice not received in the Church, negligence in studying and reading the Scriptures, negligence in rebuking vice amounting to flattery, negligence in doing everything required by his office, scurrility, lying, slander, dissolute words, injurious words, foolhardiness and evil devices, avarice and too great parsimony, undisciplined anger, quarrels and contentions, laxity either of manner or of gesture and like conduct improper to a minister. (60–61)

Notice that even these lesser offenses were not tolerated but subject to a graduated form of punishment that will be described in more detail below. If the procedure specified by Calvin was at all effective, it would be extremely difficult for a pastor to abuse his office.

The role of the elders was to "have oversight over the life of everyone, to admonish amicably those whom they see to be erring or to be living a disordered life, and, where it is required, to enjoin fraternal corrections themselves and along with others. . . . These should be so elected that there be some in every quarter of the city, to keep an eye on everybody" (63–64). Elders, like pastors, were chosen by a procedure insuring that they were acceptable to the city government in addition to those who were to be overseen. In fact, all of the elders were also members of the city government.

> In the present condition of the Church, it would be good to elect two of the Little Council, four of the Council of Sixty, and six of the Council of Two Hundred, men of good and honest life, without reproach and beyond suspicion, and above all fearing God and possessing spiritual prudence. . . . The best way of electing them seems to be this, that the Little Council suggest the nomination of the best that can be found and the most suitable; and to do this, summon the ministers to confer with them, after this they should present those whom they would commend to the Council of Two Hundred, which will approve them. If it find them worthy, let them take the special oath, whose form will be readily drawn up. And at the end of the year, let them present themselves to the Seigneury for consideration whether they ought to be continued or changed. It is inexpedient that they be changed often without cause, so long as they discharge their duty faithfully. (63–64)

It is a hallmark of Calvinism that leaders must set a high standard of morality for the body of the church. Calvin did not leave this to chance but implemented a selection and supervision procedure that made exploitation by leaders exceedingly difficult. The average Genevan might complain about the degree of commitment and discipline expected of him, but he could not complain about lapses of commitment and discipline in his leaders.

The conduct of the citizenry was enforced in an escalating fashion; first by private "brotherly admonition," then in front of the church, then by exclusion from participating in the communion of the Lord's supper (held six times a year), and finally by excommunication (this convention dates back to the early Christian church). The purpose in all cases was to accomplish a change of heart and behavior, after which the misconduct would be forgiven and the member received back into the church. The following passage from the Ordinances shows that church discipline was thorough and unforgiving of deviance, but otherwise respectful of the deviant individual:

There follows the list of persons whom the elders ought to admonish, and how one is to proceed. If there be anyone who dogmatizes against the received doctrine, conference is to be held with him. If he listen to reason, he is to be dismissed without scandal or dishonour. If he be opinionative, he is to be admonished several times, until it is seen that measures of greater severity are needed. Then he is to be interdicted from the communion of the Supper and reported to the magistrate. If anyone is negligent in coming to church, so that a noticeable contempt of the communion of the faithful is evident, or if he show himself contemptuous of the ecclesiastical order, he is to be admonished, and if he prove obedient dismissed in friendliness. If he persevere in his evil way, after being three times admonished, he is to be separated from the Church and reported. As for each man's conduct, for the correction of faults, proceedings should be in accordance with the order which our Lord commands. Secret vices are to be secretly admonished; no one is to bring his neighbour before the Church to accuse him of faults that are not in the least notorious or scandalous, unless after having found him contumacious. For the rest, those who despise particular admonitions by their neighbor are to be admonished anew by the Church; and if they will not at all come to reason or acknowledge their fault when convicted of it, they will be informed that they must abstain from the Supper until such time as they return in a better frame of mind. As for vices notorious and public which the Church cannot dissimulate, if they are faults that merit admonition only, the duty of the elders will be to summon those who are implicated to make friendly remonstrance to them in order that they make correction, and, if amendment is evident, to do them no harm. If they persevere in doing wrong, they are to be admonished repeatedly; and if even then there is no result, they are to be informed that, as despisers of God, they must abstain from the Supper until a change of life is seen in them. As for crimes which merit not merely remonstrance in words but correction by chastisement, should any fall into them, according to the needs of the case, he must be warned that he abstain for some time from the Supper, to humble himself before God and to acknowledge his fault the better. If any in contumacy or rebellion wish to intrude against the prohibition, the duty of the minister is to turn him back, since it is not permissible for him to be received at the Communion. Yet all of this should be done with such moderation, that there be no rigor by which anyone may be injured; for even the corrections are only medicines for bringing sinners to our Lord. (70–71)

In contrast to what the phrase "turn the other cheek" implies when taken out of context, Calvin's church was elaborately protected by a system of social controls designed to eliminate deviant behavior. It was highly resistant to exploitation, even by its own most powerful members. According to one Calvin scholar (Collins 1968, 148): "Neither rank, nor wealth, nor social position made one immune from being summonsed before it. Favres, Perrins, Bertheliers [powerful Genevan families] were just as subject to its censures as the most humble artisan and the evidence shows that decisions were meted out with impartial justice."

Evolutionary theories of altruism and cooperation focus almost exclusively on the problem of cheating. Even when this problem can be solved, however, formidable problems of coordination remain. We frequently talk about the infrastructure of modern society, but Geneva also had an infrastructure, and the burden of supporting it was probably greater for the average citizen then than now. A massive wall around the city had to be maintained, the Swiss mercenaries that protected Geneva from the duchy of Savoy had to be paid, a plague hospital had to be built and staffed, charity had to be given to the poor, an educational system had to be built; the list of public goods goes on and on. The temptation to avoid the burden must have been great, not to speak of subverting the entire system. Naphy (1994, 12) refers to factionalism prior to Calvinism as "the Genevan disease." Civic authority by itself was unable to forge such an unruly population into an adaptive unit, which is why the unifying effect of religion was needed. God's will for citizens of Geneva was to shoulder the burden of the city's infrastructure.

How can a large population be made to do the right things at the right time, even after they are willing? The Ecclesiastical Ordinances specified a decision-making structure, an educational system, a healthcare system, and a welfare system. Decisions were made by consensus among the pastors during weekly meetings without any formal status differences. For disagreements that could not be resolved, the decision-making circle was widened to include first the elders and ultimately the city council. This structure realized the benefits of group-level decision making (reviewed by Wilson 1997) and also made it difficult for any single individual to impose his will on the community.

The educational system was supervised by the doctors, who not only instructed all children in the catechism but also trained adults for the ministry and included scholars of the Old and New Testaments whose learning was used for "defending the Church from injury by the fault of pastors and ministers" (62). Children learned the catechism by memorizing fifty-five weekly lessons in a question-and-answer format (e.g., Minister: "What is the chief end of human life?" Child: "That men should know God by whom they were created"). Only after passing an examination were they allowed to participate in the Lord's Supper. At Calvin's urging, the city of Geneva "rose above its poverty" (McNeill 1954, 195) and created the Geneva Academy, which eventually attracted students from throughout Europe. Training was so thorough that it was said that the boys of Geneva talked like Sorbonne doctors.

The health and welfare systems were supervised by the deacons, who were chosen in the same way as the elders. An isolated plague hospital was maintained. The unemployed and destitute were given lodging, and those who became sick were placed in separate rooms. A doctor and surgeon were provided for those who could not afford their own health care. Money was raised for charity and distributed on a case-by-case basis according to need. Both the raising and spending of money were monitored by accounting procedures that made it difficult to cheat, although the deacons chosen for the task were such solid citizens that they gave in time and effort far more than they could have gained (Olson 1989).

The duty of attending to dying plague victims provides a good illustration of how the pastors solved problems in an egalitarian fashion using mechanisms that are difficult to subvert from within. This life-threatening task was decided by lottery. Calvin was exempted from the lottery by decree of the city council because his death would have a far greater impact than that of the other pastors on the fate of the church. This was the only duty that Calvin did not share equally with the other pastors. When Calvin was succeeded upon his death by Theodore Beza, Beza himself successfully lobbied the city council to be included in the lottery (Monter 1967, 210).

To summarize, Calvin's church included a code of behaviors adapted to the local environment, a belief system that powerfully motivated the code inside the mind of the believer, and a social organization that coordinated and enforced the code for leaders and followers alike.[6]

THE DARK SIDE OF CALVINISM

In chapter 1 I emphasized the moral complexity of multilevel selection theory. Darwin proposed group selection to explain the evolution of human moral virtues, but by his own account these virtues are practiced within groups and against other groups. Even within groups, the so-called moral virtues are often enforced by coercive mechanisms. Finally, what counts as moral conduct for one group may appear bizarre and highly immoral to members of other groups. An evolutionary theory of morality needs to be this complex to explain the complexity and ambiguity of moral issues in everyday life.

So far I have described within-group features of Calvinism that many modern readers would regard as morally praiseworthy. It is difficult not to admire a man such as Beza, who in his position as leader among equals asks for a fair chance to care for dying plague victims. However, it is not my purpose to paint a romantic portrait of Calvinism, which also included elements that seem bizarre and morally repugnant not only to the modern mind but also to some of Calvin's contemporaries. Let us therefore look at the dark side of Calvinism to see if it can be explained by multilevel selection theory as successfully as the bright side.

Calvin attempted to implement a high degree of social control in Geneva. Families were visited once a year to have their spiritual health examined. Church attendance was required, and playing a game of skittles on Easter Sunday was sufficient to send the son of a prominent family to prison. Some of Calvin's most famous battles with the rich and powerful of Geneva were based on such offenses. Ami Perrin, the top military commander of the city, was imprisoned for dancing inappropriately at a wedding. Unrepentant, Perrin refused to appear before the Consistory, which prompted the following letter from Calvin (quoted in Collins 1968, 157):

I beg you to reflect on this. I am unable to have two weights and two measures, and as it is necessary, as a matter of law *(en droit)* to maintain equality, one is doubtless unable to tolerate inequality in the Church of God. Who I am, you know, or ought to know: I am he to whom the law of the Celestial Heritage is so dear that no one can prevent me from defending it with a good conscience. (As for me, I desire in this matter to consult not only the edification of the Church and your salvation, but also your convenience, name and leisure, for how odious would be the imputation which is likely to fall upon you, that you were apparently free and unrestrained by the common law to which everyone is subject? It is certainly better, and in accord with my zeal and your welfare, to anticipate the danger than you should be so branded.) I am well aware of what has been said in your house, that I should take care of rekindling a smouldering fire, lest that should happen to me which occurred several years ago [being expelled from Geneva]. But such suggestions have no weight with me. For I did not return to Geneva for repose, or for gain; and I should not complain if I had again to leave Geneva. It is the interest and safety *(salut)* of the Church and of the city which have brought me here. . . . But the indignity, the ingratitude of some will not cause me to fail in my duty or to lay down, before my last breath, the zeal God has stirred up in me for this city.

Calvin ended his letter by saying "God keep you and show you how much the blows *(coups)* of a sincere friend are better than the treacherous flattery of certain others." Although the fact that this confrontation arose over a seemingly trivial offense may be puzzling to us, it is still possible to explain the nature of the confrontation in terms of multilevel selection theory. The issue was whether powerful members of the community were to be held to the same moral standards as everyone else. Calvin's identification with the welfare of the entire city and his use of religion to achieve a moral ground higher than any secular authority provides an excellent example of the basic thesis of this book.

The scrutiny of behavior that seems strange to us needs to be evaluated against the background of the times. Another conflict between Calvin and Perrin involved a clothing fashion in which one kind of cloth was sewn into slits in a garment in a way that puffed outward. Perrin wanted to outfit his troops in this fashion for an archery festival until Calvin intervened, calling the issue a matter of "superior and fundamental principles." We can begin to understand how Calvin could feel so strongly about a fashion trend when we realize that the style was regarded as indecent and had been outlawed in Augsburg, Zurich, and Berne before Calvin even arrived in Geneva (Collins 1968, 158).

Although we may find it difficult to understand what was at stake in controversies over dancing and dress, there is no mistaking the seriousness of other perceived threats. If Amnesty International existed in the sixteenth century, with its current norms, every religious and political organization in Europe would be on its list. Banishment, torture, and execution were acceptable forms of social control to punish seriously deviant behavior. We have seen that Calvinism fostered a degree of mutual criticism, differences of opinion, and the internal checks and balances that are required for a group to make intelligent decisions and to prevent subversion from within. However, dissension from outside the church elicited another kind of reaction altogether. A menacing placard placed in the pulpit of St. Peter's cathedral led to the arrest of one Jacques Gruet, a friend of the powerful Favre family. Gruet confessed under torture, and a search of his house revealed more incriminating evidence. As Collins (1968, 161–62) tells the story:

> He called Calvin "a great hypocrite" who wished to have himself adored and to have "the dignity of our holy father the pope." Moses, he asserted, said much and proved nothing, and he attributed to human caprice all laws, whether divine or human. On the margin of a book of Calvin's on immortality Gruet had written "All folly." Gruet, says Roget, was a free thinker and a liberal in the modern sense. He distinguished between offences against God, with which the magistrate need not concern himself, and offences against society, which ought to be repressed. As Choisy has pointed out, Calvin could not understand the distinction. "Society rests on the authority of God, its duty is to maintain the honor of God, to respect God as its sovereign and the will of God as sacred law."
>
> Gruet was executed for blasphemy and lese-majesty. As far as is known not one of his Libertine friends tried to help him.

The church of Geneva did not have the authority to execute anyone, so this verdict must have been made by the Genevan civil government. The lack of support for Gruet suggests that Calvin did not need to play a direct role in the decision. The emergence of religious tolerance was a future event in European history and may itself have been an important factor in the evolution of still larger adaptive social organizations. We can sympathize with Gruet from our modern perspective and still understand what took place in the context of multilevel selection theory.

Calvinism's most infamous deed was the execution of Michael Servetus (1511–53) for heresy. Calvin has often been blamed directly for this deed, but most Calvin scholars agree that he was part of a broader consensus in which, if anything, he played a moderating role. Servetus initially was welcomed by other reformers, but he ultimately alienated them with his attacks on core Christian doctrines. First Oecolampadius and then Bucer hosted and engaged him in long dialogues, only to terminate their association in exasperation. His book *The Errors of the Trinity* was banned in Strasbourg and Basel. Ultimately Servetus was obliged to leave the Rhine country for France, where he masqueraded for twenty years as a Catholic under the assumed name of Michel de Villeneuve, studying medicine and supporting himself as a physician, editor, and proofreader. In an edition of Ptolemy's *Geography* that he helped to produce, he included a comment by a Dutch scholar that Palestine, far from being a land flowing with milk and honey, was uncultivated and sterile. This innocent act would later come back to haunt him.

Behind his Catholic façade, Servetus continued to think and write with undiminished zeal and eventually initiated a dialogue with Calvin, using a publisher as a go-between. His first letter displayed the same patronizing arrogance that alienated the other reformers. Calvin did not mince words in his nineteen-page reply: "There is no honesty in you nor plain dealing worthy of a liberal spirit. . . . You are ridiculous and unreasonable." He sent a copy of the *Institutes*, which was returned annotated with insulting comments. Servetus wrote over thirty letters to Calvin before the latter called an end to the relationship. In a letter to Farel he wrote "Servetus has recently written to me and to his letters has joined a long volume of his ravings. . . . If I consent he will come here. But I do not wish to pledge my word. For if he comes and my authority is worth anything I shall not suffer him to leave alive."

Servetus managed to find a publisher for his "ravings," which quickly led to his imprisonment in France. A Genevan publisher and friend of Calvin was instrumental in his conviction, using some of Calvin's letters as evidence. Calvin's direct role in the affair is unclear. In any case, Servetus escaped from prison and turned up in Geneva four months later, where he was recognized and again imprisoned. It is not known why Servetus traveled to Geneva; one theory is that he was plotting with Calvin's opponents for his overthrow. In any case, in the ensuing trial Servetus allied himself as closely as possible with Calvin's powerful enemies, who in turn saw their chance to attack Calvin on his own theological turf. The trial seemed to hinge on theological issues such as the doctrine of the Trinity, infant baptism, and the scandalous suggestion that Palestine was not a land flowing with milk and honey. In reality it was a battle for the moral high ground, the only basis for authority that Calvin and his church ever had.

Just as the decision-making circle within Calvin's church was widened for the most controversial issues, the verdict for the trial of Servetus was ultimately referred to Protestant churches outside the city of Geneva. All of them condemned Servetus, although without specifying the nature of the punishment. Calvin argued for a less cruel form of execution, but the majority decision was for Servetus to be burned at the stake. Calvin certainly played a role in the death of Servetus, but it is a gross distortion to suggest that he was solely responsible or that he imposed his will upon others against the background of his times.

Two years later, in 1555, Perrin's opposition to the church erupted in a melee that led to his expulsion and the execution of four of his associates. The executioner botched his job and caused a more agonizing death than intended, prompting Calvin to observe that their extra suffering may have been God's will. Calvin was not a paragon of virtue, but his moral failings occurred in exactly the contexts predicted by multilevel selection theory: social control within groups and conduct toward members of other groups.

Taking Stock

Science is a feedback process between hypothesis formation and testing. Apart from the brief examples that I presented in chapters 1 and 2, this is our first attempt to test the hypothesis that religious groups can qualify as adaptive units, by looking in detail at a single religious community in relation to its environment. Did Calvin's church possess the kind of adaptive complexity implicit in the word organism, or it is better explained by the other major hypotheses outlined in table 1.1? How did group-level functionality arise, if indeed it exists? More generally, does an evolutionary perspective add anything to our understanding of Calvinism, which has already been so well studied from other perspectives?

The argument from design

In chapter 2 I reviewed the methods that evolutionary biologists use to study adaptations in nonhuman species. One of these is the argument from design. If an object or an organism is adapted to perform a given function, it must have certain design features that are unlikely to have arisen by chance. There are many ways to be nonadaptively complex but only a few ways to be adaptively complex. Like the concept of adaptation itself, the argument from design is much older than Darwin's theory of evolution. Paley (1805) used the analogy of a watch proving the existence of a watchmaker to argue that the intricate structure and function of organisms proves the existence of God. If we replace "God" with "designing agent," the argument remains perfectly sound when used with appropriate care (Sober 2001).

Under the banner of "just-so story!" academic discussions have made the argument from design appear more tenuous than it really is and have obscured the extent to which it is used in everyday life. Suppose I ask you to identify an object that I place in your hand. You glance at it, look at me quizzically, and reply "It's a can opener." I ask if you're sure it isn't an ax for chopping wood or a mineral deposit dug from the earth. You give me a long, strange look, wondering if I am playing a trick or if I have all my marbles, before replying "No, it's a can opener." When I ask you to justify your conclusion you speak in slow, measured syllables: "Well, this smooth-edged wheel is used to pierce the metal, this tooth-edged wheel is used to grip the rim of the can, this lever applies the force that pierces the metal, and this knob allows one to rotate the wheels to move the can opener around the rim. It is incapable of chopping wood and no mineral deposit has ever been dug from the earth with moving parts that articulate with each other in just the right way to open cans."

Clearly, the function of a can opener is so painfully obvious that this conversation could only take place with a joker or an insane person. The design features that identify the object as a can opener provide such a strong argument that we don't even call it an argument; we call it self-evident. We use the argument from design so often and reflexively in our everyday lives that we would be paralyzed without it. I do not mean to underestimate the difficulty of inferring function, but academic discussions have erred in the opposite direction.

What makes the argument from design so compelling in the case of the can opener is detail. The more we know about the various parts of the object in relation to each other and their environment (the can), the more obvious their functional nature becomes. Similarly, the absence of function becomes more obvious with every new detail for objects that truly have no function. A cloud or an ice crystal will never be shown to have a utilitarian function, no matter how long they are studied. All of these principles apply with equal force to biological adaptations, allowing the functions of organs such as the heart and eye to be established long before Darwin's theory of evolution.

Our task is to use the same reasoning process to infer the function—if any—of a religion such as Calvinism. As with the can opener or the heart, the key is to understand the details. This point is important because my analysis of Calvinism may not have appeared very scientific. It was purely descriptive, without a single number or statistical comparison. Of course, I have nothing against numbers and statistical comparisons, but it is important to see them as a refinement of something more basic. I mentioned earlier that traditional religious scholarship is comparable to the information from natural history that was sufficient for Darwin to establish his theory of evolution. To emphasize this point, consider what we would gain by making the study of Calvinism more quantitative. According to Collins (1968, 148; quoted above), "the evidence shows that decisions were meted out with impartial justice." The evidence to which Collins refers consists of decisions that are preserved in the historical record, in which the high and mighty seem to be as accountable as the average citizen. Collins could be wrong, especially if other Calvin scholars who have sifted through the same material disagree with him. In this case, it might be necessary to resolve the issue with more rigorous methods. We might assemble all the disciplinary actions that we can find into a data base and develop numerical scores for social status and the outcome of the judicial decision. Then we could ask statistically whether the average status of people disciplined by the church differed from the average status of people in the city of Geneva and whether status influenced the outcome of the decision for those who were judged. I can easily imagine that statistically significant differences would be found, despite the high-profile cases that I have described above and that caused Collins to reach his conclusion. However, to go further we should extend the analysis to other churches, especially the Catholic Church that preceded Calvinism in the city of Geneva. Even if Calvinism wasn't perfectly fair in its administration of justice, was it more fair than what it replaced?

This kind of quantitative analysis is always welcome and is reported in academic publications such as the *Journal for the Scientific Study of Religion*, although not for the subject of Calvinism as far as I can tell. I will survey this literature in chapter 5, but for the moment it needs to be seen as a refinement of what careful religious scholars have been doing all along. Furthermore, it is a labor-intensive process that must be reserved for the least certain issues. There is a such a consensus among historians that the

Catholic Church in Geneva was corrupt and beholden to vested interests outside Geneva in comparison to Calvinism that there seems little point in toting up the numbers, any more than we need to quantify the function of a can opener. If we ignore the general scholarly literature as "unscientific," we risk throwing out the self-evident in addition to the uncertain. Instead, we should treat the scholarly literature as a source of reasonably accurate information and proceed to test our hypotheses. I claim that knowledge of the details clearly supports a group-level functional interpretation of Calvinism. Calvinism is an interlocking system with a purpose: to unify and coordinate a population of people to achieve a common set of goals by collective action. The goals may be difficult to define precisely, but they certainly include what Durkheim referred to as secular utility— the basic goods and services that all people need and want, inside and outside of religion. The interlocking system includes explicit behavioral prescriptions, specific theological beliefs, and a mighty fortress of social control and coordination mechanisms. Thinking of a group as an organism encourages us to look for adaptive complexity, and this search has paid off in the case of Calvinism. Refining the information by making it quantitative might change some of the details, but I would be very surprised if it changed the fundamental conclusion.

Consider the alternative hypotheses: Stark's (1999) theory of religion assumes that people are cost-benefit reasoners who invent supernatural agents to provide goods and services that in fact cannot be had. I have genuine respect for the research program of Stark and his colleagues, but when his list of propositions is taken at face value it says virtually nothing about the fundamental problem of social life or the role of religion in its solution. As far as I can tell, Calvinism is no better explained as the economic mind spinning its wheels to get what it can't have than a can opener is explained as a device for chopping wood. Unless Calvinism is highly unrepresentative of other religions, we should be growing suspicious of Stark's ambitious claim to explain every nuance of religion on the basis of his formal theory.

Or consider the alternative hypothesis that Calvinism can be explained on the basis of secular utility, but at the level of the individual rather than the group. To avoid the pitfalls of the averaging fallacy discussed in chapter 1, we must be careful about how we make our comparisons. The question is whether Calvinism is designed to benefit some individuals (presumably the leaders) at the expense of others within the church, or whether Calvinism is designed to benefit everyone within the church as a collective unit. Exploitation is sometimes naked, but more often it requires deception. Thus, we might expect Calvinism to appear good for the group at first sight but to emerge as a tool of exploitation upon closer examination. Some features of some religions undoubtedly can be explained in this way, such as many practices of the Catholic Church that led to the Reformation. Religious scholars have already reached this conclusion on the basis of qualitative information, and there is little reason to challenge it. I am not trying to explain every feature of every religion as good for the group, nor could I succeed if I tried. It is multilevel selection theory that explains the nature of religion, not group selection alone. Churches *are* subverted from within, which in part accounts for the formation of new churches, as I will describe in more detail in chapter 5. However, the details of Calvinism point strongly to genuine group-level benefits rather than an elaborate sting operation. Calvin's age was one of verbal hyperbole and outrageous accusations. Some of the bitter enemies of Calvinism accused the pastors of enriching themselves, but this claim has been dismissed as laughable by every Calvin scholar that I have encountered. Calvin's passion for setting a moral example of selflessness is well known, culminating in his burial in an unmarked grave, but more important are the many features of the religion that are clearly designed to constrain the self-will of the leaders as effectively as the rank-and-file. Once we avoid bloated definitions of self-interest as "anything that works at any level," Calvinism can be explained on the basis of group-level benefits with little need for controversy. Without claiming too much for the power of the argument from design, the details of Calvinism are better explained as a group-level adaptation than by any other competing hypothesis. A can opener is designed to open cans, and Calvinism is designed to make a human community function as an adaptive unit.[7]

Innate psychology and cultural evolution

In chapter 1 I showed that human moral systems have both a genetically evolved component and an open-ended cultural component. An innate psychological architecture is required to have a moral system, but the specific contents can vary and therefore adapt to recent environments. In addition, it is important to remember that moral communities larger than a few hundred individuals are "unnatural" as far as genetic evolution is concerned because to the best of our knowledge they never existed prior to the advent of agriculture. This means that culturally evolved mechanisms are absolutely required for human society to hang together above the level of face-to-face groups.

It is easy to recognize in Calvinism the same kind of guarded egalitarianism that characterizes hunter-gatherer groups, with cultural add-ons that allow the mechanisms to work at a larger scale. It is impossible for everyone to monitor the behavior of everyone else in a city the size of Geneva, but elders can be elected to supervise sections of the city whose inhabitants they do know well. Thousands of people cannot participate in a consensus decision-making process, but a group of pastors is much like a group of hunter-gatherers squatting around a campfire. Thousands of people do not converge naturally on shared norms of behavior, but uniformity can be achieved by the many devices employed by Calvinism and other religious systems. Above all, Calvinism shares with hunter-gatherer morality a fundamental distrust of human nature, a distrust that expects and guards against exploitation at all levels of the social hierarchy it creates. The emerging paradigm of major transitions in biology is centered on the concept of social control mechanisms that prevent subversion from within, converting groups of organisms to groups as organisms. We saw in chapter 1 that this paradigm can be extended to human genetic evolution and hunter-gatherer groups. The example of Calvinism shows that it can also be extended to human cultural evolution and the nature of modern societies.

Jacob (1977) described evolution as a tinkerer that builds new structures out of old parts. The tinkerer metaphor aptly describes how innate psychological mechanisms do not necessary limit cultural evolution but rather provide the building blocks that cultural evolution uses to create innumerable forms. Forgiveness and faith are two examples of capacities that are part of the psychological toolkit of all normal humans and that have obvious functions outside the context of religion, which are put to new use by a culturally constructed religious belief system. Calvinism's grip on basic human psychology has been noted by Calvin scholars such as McGrath (1990, 224), who identified "the fundamental contribution of Calvinism as lying in its generation of psychological impulses on account of its belief systems." McGrath attributed this insight to Max Weber (1930), who famously proposed that Calvinism gave rise to the Protestant work ethic and the spirit of capitalism that now pervades modern life. Weber's thesis illustrates the other side of the tinkerer metaphor—not the old parts, but the new structure that can be explained only in terms of modern history and by no means as an adaptation to ancestral environments. Weber's thesis is still taken seriously by social scientists, as well it should be (e.g., Lessnoff 1994). Human nature may be built from old parts, but it still can evolve by ongoing cultural evolutionary processes.

What are the designing agents?

In chapter 2 I stressed that functional design can be demonstrated without reference to the designing agent(s). William Harvey ([1628] 1995) was right about the function of the heart and circulatory system but wrong about the creator as the designing agent. The Chicago economists identified the functional nature of business firms without knowing whether the designing agent(s) were rational thought, blind imitation of success, or an even blinder process of differential survival and reproduction of firms. It is important to separate the question of design from the question of the designing agent(s), which is why I think that Elster's concept of intentional explanation should be regarded as a type of functional explanation. In the case of Calvinism, we can conclude that it is an interlocking system with a purpose before we ask about the designing agent(s). Once function has been established, however, we clearly want to know how it came to exist.

One possibility is that Calvin and others consciously intended to create a religious system with a purpose and planned it in every detail. There is little doubt that conscious intentionality can explain some of the functionality of Calvinism and other religions. Calvin and his contemporaries were in part hard-headed realists who talked explicitly about the ingredients required to hold a community together. However, it is unlikely that conscious intentionality can explain all of the functionality of Calvinism or other religions. For example, Calvin evidently believed to the depth of his soul that Christ's return to earth was imminent, requiring urgent preparation on the part of the church. Similar beliefs in imminent events have arisen repeatedly in religions around the world and have a clear latent function (motivating action) that differs from their manifest function. In general, there is something absurd about trying to explain the functional properties of religion exclusively in terms of the conscious intentions of its framers, when the whole problem, as Durkheim realized, is that religious belief appears so otherworldly that it must be explained by latent functions if functions exist at all. We can safely conclude that conscious intentional thought explains some but not all of the functionality of Calvinism. The beliefs that motivate religious people to behave as they do in their own minds (the manifest functions) often depart from the adaptive consequences that ultimately sustain the beliefs (the latent functions).

Conscious intentional thought is only the latest gadget in a toolkit of psychological mechanisms that evolved to transform environmental information into adaptive behavior, many of which operate beneath conscious awareness. These mechanisms may also contribute to the group-level functional design of a religious system such as Calvinism. Whenever the designing agent can be traced to a psychological mechanism (conscious or otherwise), we must ask why that mechanism exists, compared to many others that can be imagined. For example, suppose that Calvin was guided by an imitation instinct that caused him to look around and copy the most successful practices of other churches. Why should this instinct exist, compared to many others that can be imagined? In particular, why didn't he copy the most successful individual in Geneva, rather than working so hard to make Geneva copy the most successful group? Whenever group-level functional design is traced to a psychological mechanism, we have not avoided the topic of multilevel selection but moved it backward to the origin of the psychological mechanism. Barring theological explanations, all designing agents must ultimately be traced back to a process of blind variation and selective retention (Campbell 1960).

In addition to providing the ultimate explanation for psychological designing agents (conscious or unconscious), multilevel selection might also have contributed directly to the adaptive design of Calvinism. The Protestant Reformation can be regarded as a large number of social experiments with many failures for each success. The Swiss reformer Zwingli was arguably more gifted and charismatic than Calvin, but his effort to create a church in Zurich ended in disaster. The factors that cause one social experiment to succeed while others fail are probably so complex and historically contingent that they will never be fully understood, and they certainly extend beyond the conscious intentions of the individual actors. Nevertheless, the success of some and the failure of others on the basis of their respective properties constitutes a process of blind variation and selective retention operating during the course of recent human history in addition to the distant evolutionary past.[8] Ongoing cultural evolution, unconscious psychological mechanisms at both the individual and group levels, and conscious intentional thought must all be considered as potential designing agents for a modern religious belief system such as Calvinism.

RATIONALITY, RELIGION, AND ADAPTATION

As I mentioned earlier, people who stand outside of religion often regard its seemingly irrational nature as more interesting and important to explain than its communal nature. Rational thought is treated as the gold standard against which religious belief is found so wanting that it becomes well-nigh inexplicable. Evolution causes us to think about the subject in a completely different way. Adaptation becomes the gold standard against which rational thought must be measured alongside other modes of thought. In a single stroke, rational thought becomes necessary but not sufficient to explain the length and breadth of human mentality, and the so-called irrational features of religion can be studied respectfully as potential adaptations in their own right rather than as idiot relatives of rational thought.

The seemingly irrational features of Calvinism seem gratifyingly functional from an evolutionary perspective. For all its otherworldliness, Calvinism caused its community of believers to behave adaptively in the real world, which is all that evolution can be expected to accomplish. Calvinism even provides a natural before-and-after experiment on the effect of religious belief on group-level functional organization. Before Calvin, the city of Geneva was riven by internal discord despite a strong civic government and an impelling need to pull together. When Calvin and Farel first proposed their plans for Geneva they were promptly expelled, but they were asked to return three years later because the city feared for its survival as a social unit:

> Events in the absence of Farel and Calvin had demonstrated the close interdependence of reformation and autonomy, of morals and morale. Although the city council was concerned primarily with the independence and morale of the city, the fact that Farel's religious agenda could not be evaded gradually dawned. The pro-Farel party probably had little enthusiasm for religious reformation or the enforcement of public morals; nevertheless, it seemed that the survival of the Genevan republic hinged upon them. (McGrath 1990, 103)

McGrath's juxtaposition of the words "morale" and "moral," and their close etymological connection, suggests that a strong and shared sense of morality is required to galvanize and orient a human community toward a common purpose—for modern societies no less than for hunter-gatherer groups. The effect of Calvinism on Geneva was so profound that the city assumed an importance in world affairs out of all proportion to its economic significance:

> Notwithstanding the repressive discipline, harsh laws, and paternalistic controls, the positive and constructive elements of Calvin's system were becoming more and more effective. The people of Geneva listened to

preaching several times weekly. A new generation was arising, trained in Calvin's Sunday School, instructed by his sermons, able to recite his catechism, to sing the Psalter, and to read the Bible with understanding. Possibly no community had ever before existed so well indoctrinated and broken to discipline. (McNeill 1954, 190)

In the future it will be important to quantify the effect of Calvinism on the functional organization of Geneva, but the numbers will probably broadly confirm what Calvin scholars have already concluded. This natural before-and-after experiment provides powerful support for the organismic view of religion, support as powerful as can be expected from a single case study.

When the great anthropologist E. E. Evans-Pritchard (1956, 314) reviewed the theories of religion of his day, he came to the following pessimistic conclusion:

> It is indeed surprising that those writers, whose speculations were for the most part attempts to explain religion as a general phenomenon, should have turned their attention exclusively to the religions of present-day primitive peoples or to the earliest forms of the higher religions—to those religions, that is, about which information was the most lacking and the most uncertain—rather than to the contemporaneous world religions with their vast literatures and known histories, to Christianity, Islam, Hinduism, Buddhism, and others. Had they done so, however, and, even more, had they conducted research into what these religions mean to ordinary people rather than into how philosophers, theologians, lawyers, mystics, and others have presented them, they would have seen how inadequate their theories were.

When a scientific hypothesis is on the right track, it is strengthened rather than weakened by additional information, even if minor amendments are required along the way. I would like to think that the organismic hypothesis has survived Evans-Pritchard's test, becoming stronger on the basis of the enormous amount of information available for a modern religious system such as Calvinism. Let us now broaden our scope to include the religions that have existed throughout time and around the world.

CHAPTER 4

THE SECULAR UTILITY OF RELIGION
Historical Examples

We spent many hours in Cwm Idwal, examining all the rocks with extreme care, as Sedgwick was anxious to find fossils in them; but neither
of us saw a trace of the wonderful glacial phenomena all around us; we
did not notice the plainly scored rocks, the perched boulders, the lateral
and terminal moraines. Yet these phenomena are so conspicuous that . . .
a house burnt down by fire did not tell its story more plainly than did
this valley. If it had still been filled by a glacier, the phenomena would
have been less distinct than they are now.
                                                   —Darwin [1887] 1958, 70

This passage from Darwin's autobiography wonderfully illustrates how
a theory is required to see the things that are in front of our faces. The
young Darwin and his professor Adam Sedgwick could not see the evidence for glaciers because the theory of glaciation had not yet been proposed. With the theory in mind, the confirming evidence was so obvious
that a glacier itself might as well have still been present.

A similar obviousness in retrospect may exist for the organismic concept of religious groups. If it is true in a strong sense, then the evidence
must be in front of our faces, like the boulders and moraines that Darwin
wandered among in search of fossils. We fail to see the evidence, not
because it is obscure or requires sophisticated measuring devices, but because we are employing the wrong theories. Granted, the theory that
explains religious groups as adaptive units is not new, as the theory of
glaciation was in Darwin's day. Nevertheless, the demise of group selection
in biology and functionalism in the social sciences made the evidence
seem to disappear, as surely as if these theories had never been proposed.[1]

Now that the organismic concept of groups has become theoretically robust, we need to take a fresh look at what has always been in front of our faces.

In the last chapter I took a fresh look at Calvinism from an organismic perspective. In this chapter I will expand my scope to include religions from across time and around the world. First I will review three examples that provide exceptionally clear evidence for what Durkheim called the secular utility of religion. Then, to avoid the possibility that these are a biased sample, I will describe the initial stages of a survey of twenty-five religious systems chosen at random from the sixteen-volume *Encyclopedia of World Religions* (Eliade 1987). In the next chapter I will review the modern social scientific literature on religion.

EXAMPLE 1: THE WATER TEMPLE SYSTEM OF BALI

At the summit of a volcano on the island of Bali stands an immense temple for the worship of Dewi Danu, goddess of the waters that are thought to emanate from the volcano's crater lake. The temple is inhabited by twenty-four priests, chosen as children by a virgin priestess to be lifelong servants of the goddess. The high priest is called the Jero Gde and is thought to be an earthly representative of the goddess herself. The Jero Gde is always dressed in white and wears his hair long. By day he offers sacrifices on behalf of the farmers who rely on water to irrigate their rice crops. By night he receives guidance from the goddess in his dreams. Although the Jero Gde is chosen as a child from a commoner caste, his status rivals that of the most powerful kings.

The rain that falls on the mountain tumbles down to the sea in rivers that cut deeply into the soft volcanic rock. To use the water for irrigation, the Balinese have created a vast system of aqueducts, often running through tunnels a kilometer or more in length, that shunt the water from the rivers to the rice terraces, also sculpted by human labor from the steep mountain slopes. Below the grand temple of the crater lake stand smaller temples at every branch of the irrigation system, ending with the smallest temples where the channels empty onto the fields. These temples stand empty except for appointed days, but each has its own pantheon of deities. In addition to the water temples there are other temples, large and small; in houses, market places, villages, palaces, throughout the island of Bali, whose deities are drawn from the original indigenous religion and from the more recent religions that arrived by trade, migration, and invasion.

This description of Balinese religion reads like a scene from a Holly-wood movie. In academic terms it seems far more likely to be explained as an accretion of history and a costly byproduct of human mentality than as the adaptive physiology of a group-level organism. Nevertheless, anthropologist Steve Lansing and his colleagues have demonstrated a profound functionality that lies at the heart of the water temple system of Bali (Lansing 1991; all subsequent page numbers refer to this book).

One thing is certain: rice cultivation in Bali reflects the coordinated effort of thousands of people over a period of many generations.[2] The Balinese landscape is so breathtaking in part because it is so highly modified by human activity. An aerial view of the rice terraces bears an eerie resemblance to the cells of a beehive or a termite colony. The smallest social unit responsible for this miracle of pre-industrial engineering is the subak, an association of farmers that shares the water emerging from a terminal branch of the irrigation system. A subak is roughly the size of a hunter-gatherer group, and its members operate in a democratic fashion, making decisions by consensus and electing their leader from among their own ranks. Lansing, quoting a member of the Dutch colonial government in Bali during the late nineteenth century, described the activities of the subaks this way:

> "The most important features of the work performed collectively are the maintenance of the dams and conduits of the subak irrigation system and the patrolling of the conduits to safeguard against theft of water. The extent to which each individual member is obliged to participate in the collective labor is assessed according to the distribution of water in the subak." This work includes (1) maintenance of the dam and conduits and the various installations; (2) inspection and patrolling of the conduits; (3) maintenance of roads and culverts; (4) policing the subak and assisting the klian (subak head); (5) construction and maintenance of buildings. . . . Liefrinck described the subaks as efficient, democratic organizations that managed irrigation more effectively in the absence of royal interference. (27)

This description nicely fits the image of hunter-gatherer egalitarianism described in chapter 1, which comes naturally to small human groups whose members are capable of policing each other. However, the Balinese irrigation system also requires coordination of effort and policing at a larger scale. Who is to build and maintain the higher branches of the irrigation system that feed water to more than one subak? What is to prevent the upstream subaks from taking more than their fair share of water from the downstream subaks?

Pest control poses another problem that must be solved at a large spatial scale. Continuous rice production leads to the build-up of numerous pest species, including rats, insects, and plant pathogens. The best control method is to alternate rice production with fallow periods, during which fields are either burned or flooded (depending on the pest) and alternative crops are grown. However, pest control often must take place at a spatial scale that includes many subaks, or else the pests will simply move from a fallow area to an adjacent area under production. How are pests monitored and decisions regarding control made and enforced at such a large spatial scale?

Social theorists such as Karl Marx and colonial rulers such as the Dutch assumed that a strong centralized political authority was required to accomplish coordination and policing at such a large scale. This assumption is almost certainly false in the case of Bali, whose agricultural system functioned largely independently of its kingdoms. Instead, rice production in Bali is orchestrated by the system of water temples and their many deities that superficially appeared so difficult to explain in functional terms. An irrigation system by its nature serves a nested hierarchy of groups. As we follow a terminal branch upstream from a single subak, it will join another terminal branch that combines the interests of two subaks. Continuing upstream, the branch combines the interests of four subaks and so on, up to the entire watershed. Each branch of the irrigation system defines a community of farmers downstream whose fate is joined by their shared use of the water. The downstream community is the congregation of the temple located at every branch. The reason that every temple has its own set of deities is because every congregation has its own unique set of coordination and policing problems, distinct from those of every other congregation. The situation is more complex than I have described because the irrigation system resembles a network more than a one-way flow down bifurcating branches, and pest-control problems join the interests of the subaks in a different way than water distribution problems. However, the more complex picture that Lansing describes demonstrates a degree of functionality even more impressive than my simplified account.

The Balinese water temple system combines a practical dimension and a metaphysical dimension, much like Calvinism, although of course all the details are different. The practical dimension is obvious from the way the Balinese describe it themselves, as illustrated by the following account from a farmer and village head:

The Pura Er Jeruk is the largest temple hereabouts, that is, the temple whose congregation includes all the farmers of the village of Sukawati. Now below this temple there are also smaller temples, which are special places of worship for the subaks—each subak has its own. There are fourteen of these temples, fourteen subaks all of which meet together as one here. They meet at the Temple Er Jeruk. Every decision, every rule concerning planting seasons and so forth, is always discussed here. Then, after the meeting here, decisions are carried down to each subak. The subaks each call all their members together: "In accord with the meetings we held at the Temple Er Jeruk, we must fix our planting dates, beginning on day one through day ten." For example, first subak Sango plants, then subak Somi, beginning from day ten through day twenty. Thus it is arranged, in accordance with water and Padewasan—that is, the best times to plant. Because here time controls everything. If there are many rodents and we go ahead and plant rice, obviously we'll get a miserable harvest. So we organize things like this: when the rodent population is large, we see to it that we don't plant things they can eat, so that they will all die—I mean, actually, that their numbers will be greatly reduced pretty quickly. (44)

This passage demonstrates a sophisticated conscious knowledge of rice production and a decision-making ability that we associate with rational thought, with none of the hocus-pocus superficially associated with religion. Higher up the mountain, the Jero Gde and his staff also function on a purely practical level as conflict resolution and irrigation system specialists. Lansing (80) recounts one example in which the Jero Gde resolved a dispute between a downstream Subak that destroyed the dam of an upstream subak to increase their share of water. Another example involved a village that wanted to start a new subak by building a channel from a newly discovered spring (80–81). The temple priests inspected the spring before granting permission to see if the project was feasible and if it would reduce water flow to the existing subaks. They then provided practical advice for the construction of the new irrigation system and fields.

Lansing marshals three forms of evidence to show that the water temple system functions adaptively to coordinate the activities of thousands of farmers over an area of hundreds of square kilometers. First, the Dutch, who conquered Bali in the nineteenth century and whose own country is a miracle of hydraulic engineering, could find little to improve in Balinese rice production other than to tax it. Second, in more recent times the so-called "green revolution" was a disastrous large-scale experiment that proves the wisdom of traditional Balinese farming methods. The green revolution involved the promotion of fast-growing and high-yielding rice varieties that require a high input of fertilizers and pesticides. Balinese farmers were required by law to grow these varieties on a continuous basis as a cash crop. The resulting explosion of pests and inequitable distribution of water created a crisis that no amount of modern technology, fertilizer, pesticide, and new resistant plant varieties could solve. Third, Lansing collaborated with an ecosystem ecologist named James Kramer to build a detailed computer simulation model which showed that the water temple system was close to optimal at solving the trade-off between water use and pest control. The combined evidence is so compelling that even the Asian Development Bank, which was responsible for promoting the green revolution, was in 1988 forced to conclude:

> The substitution of the "high technology and bureaucratic" solution in
> the event proved counter-productive, and was the major factor behind the
> yield and cropped areas declines experienced between 1982 and 1985.
> . . . The cost of the lack of appreciation of the merits of the traditional re-
> gime has been high. Project experience highlights the fact that the irri-
> gated rice terraces of Bali form a complex artificial ecosystem which has
> been recognized locally over centuries. (Quoted in Lansing 1991, 124)

The secular utility of the water temple system is so impressive, at least in retrospect, that it becomes easy to lose sight of the system's metaphysical side. Nevertheless, the metaphysical side is also so impressive that it completely obscured the practical side—for the nineteenth-century Dutch, the twentieth-century architects of the green revolution, and most academic social theorists before Lansing. As one American engineer told Lansing (115), "These people don't need a high priest, they need a hydrologist!" Inside and outside the ivory tower, it was regarded as simply inconceivable that so much practical wisdom could be embodied in a religion.

Lansing's analysis of the metaphysical side of the water temple system is as instructive as his analysis of the practical side. In addition to the temples and deities, there is an important concept of holy water that symbolically represents the interdependence of the social units that interact with each other. As with Calvinism, these metaphysical elements are not a veneer but evidently are required for the system to work in a practical sense. Religious belief gives an authority to the system that it would not have as a purely secular institution (Rappaport 1979). In general terms, this authority is stated in a manuscript kept at the temple of the crater lake: "Because the Goddess makes the waters flow, those who do not follow her laws may not possess her rice terraces" (Lansing 1991, 73). The same principle is used in more specific ways to collect taxes from each village.[3] In the minds of Balinese farmers, the authority of religion appears to be experienced as sincere belief, as Lansing discovered in the following conversation with a subak head:

> Lansing: Where does the authority of the Jero Gde come from?
> Subak Head: Belief . . . overflowing belief. Concerning Batur temple—really
> that is the center, the origin of waters, you see. At this moment, the
> Jero Gde holds all this in his hands. At the temple of Lake Batur. (77)

The water temple system of Bali perfectly illustrates the general theme of this book because it combines a religion that is extravagantly other-worldly with one of the most basic human activities required for survival and reproduction—the acquisition of food. If this isn't what Durkheim meant by the secular utility of religion, what is? In addition, the water temple system provides insight on some of the major conceptual themes developed in earlier chapters. In chapter 1, I discussed the fact that groups must be defined separately for each trait. Organisms, in the strong sense of the word, are a group of elements that behave adaptively with respect to many traits, but a group can also be organismic with respect to some traits (e.g., predator defense) and not others (resource conservation). Human social groups span the same continuum; sometimes they are all-encompassing, but usually they are more narrowly defined, and in addition individuals function as members of many groups. The water temple system is a remarkable adaptation to multiple group membership that involves a separate church for each grouping, complete with its own congregation, deities, and obligations, but within a larger religious system that adaptively relates the groups to each other.

In chapter 2, I reviewed Rodney Stark's (1999) list of propositions about religion based on rational choice theory. Proposition 9 states that the greater number of gods worshipped by a group, the lower the price of exchanging with each. This proposition fails miserably for the water temple system of Bali. The reason is not far to seek; Stark assumed that multiple gods compete with each other by offering the same service to a group of religious consumers, which should drive down prices according to the laws of supply and demand. This assumption is violated for the water temple system because each deity provides a separate service. Thus, the water temple system does not violate rational choice and economic theory in any general sense, but it does show how they go wrong when they ignore group-level functionalism.

Throughout this book I have tried to emphasize both the power of the scientific method and also the many factors that impede it for a subject such as religion. The organismic concept of religious groups can be stated as a hypothesis and tested against alternative hypotheses, just as for any other subject in science. The tests might be simple to perform and the results might even appear obvious in retrospect. Nevertheless, religion has been studied for over a century by scholars earnestly trying to employ the scientific method, without this kind of clarity of outcome. Lansing's study of the water temple system of Bali only deepens the mystery. It is a jewel of scientific reasoning and methodology, providing powerful support for the organismic concept, yet in the decade since its publication it has had no impact at all on the general debate over group selection in biology or functionalism in the social sciences.

The problem is not one of poor scholarship. On the contrary, Lansing's scholarship is exceptionally deep, and he definitely sees his own work in the context of general social theory. Neither can the problem be attributed to lack of recognition. His book won a prestigious award in the field of anthropology (the J. I. Staley Prize in 1995) and the afterward written by Valerio Valeri is so reverential, even awestruck, that Lansing clearly *proved* something to his tough-minded and normally contentious colleagues. However, Lansing discussed his work in the context of Marxist theory and the theme of relativism associated with authors such as Michel Foucault and Clifford Geertz. The term functionalism is not used (although Marxism is a form of functionalism, according to Cohen 1994), Durkheim is mentioned only in passing, and evolution is mentioned only in the sense that Marx used it to describe progressive cultural change.

My point is not to criticize Lansing for a lack of scope, much less to criticize the particular issues that he addressed. Indeed, after seeing how the nineteenth-century Dutch and twentieth-century architects of the green revolution misunderstood the nature of Balinese religion and society, one begins to appreciate why relativism is such a strong intellectual tradition within the field of anthropology.[4] Instead, my point is to reveal an intellectual landscape that has impeded scientific progress in the past but need not impede progress in the future. In many ways, scientists and other intellectuals are like the Dutch and the architects of the green revolution, who can't see what is in front of their faces because of what is in back of their faces—the ideas that organize perception. What anthropologists appreciate about cultural relativism holds with equal force in the intellectual and scientific realm. However, relativism, properly understood, is not an argument against realism or the scientific method. The nineteenth-century Dutch and twentieth-century architects of the green revolution were wrong about the water temple system, Lansing is far more right, and it is possible to prove these assertions to a reasonable skeptic's satisfaction. There *are* certain things in front of our faces, and our challenge is to discover the theoretical framework that makes them visible. The water temple system provides evidence for the organismic nature of religious groups that can be seen clearly with the proper theoretical perspective.[5]

EXAMPLE 2: JUDAISM

Judaism is among the most influential of the world's religions, both in its own right and through its influence on Christianity and Islam. Judaism has already received the attention of evolutionary biologists interested in human behavior, in part because few cultures have maintained their identity for so long or left such a detailed written record of their laws, beliefs, and history (Hartung 1995a, b; MacDonald 1994, 1998a, b). My own analysis of Judaism will examine the entire religious tradition rather than a particular version that operated in a single time and place. This increase in scale should not obscure the great heterogeneity that exists within Judaism, just as within other major religious traditions.

I have already commented that religious injunctions such as the Ten Commandments and the Golden Rule are clearly adaptive at the group level. It is almost embarrassingly obvious that groups who obey these rules will function well as adaptive units, compared to groups that do not. The more one learns about Judaism, the more this impression is confirmed. The Ten Commandments are the tip of an iceberg of commandments that, at least in their intent, regulate the behavior of group members in minute detail. It should come as no surprise that Durkheim, as the son of a rabbi, should emphasize the secular utility of religion.[6]

Two facts stand out about what the People of Israel, as depicted in the Hebrew Bible, were instructed to do by their religion. First, they were instructed to be fruitful and multiply. Their religion told them to be biologically successful. Perhaps cultural evolution strays from biological evolution in other cases, but not in this case. Second, the People of Israel were provided with two sets of instructions, one for conduct among themselves and another for conduct toward members of other groups. That is the basic concept of the covenant between God and Abraham. Toward each other, the People of Israel were expected to practice the charity and collective action that we typically associate with JudaeoChristian morality, as illustrated by the following passage from Deuteronomy (15:7–10):

... do not be hard-hearted and tight-fisted toward your needy neighbor. You should rather open your hand, willingly lending enough to meet the need, whatever it may be. Be careful that you do not entertain a mean thought, thinking, "The seventh year, the year of remission [of debts] is near," and therefore view your needy neighbor with hostility and give nothing; your neighbor might cry to the LORD against you, and you would incur guilt. Give liberally and be ungrudging when you do so, for on this account the LORD your God will bless you in all your work and in all that you undertake.

Conduct expected toward members of other groups was highly context-sensitive. Deuteronomy 24:21–22 states: "When you gather the grapes of your vineyard, do not glean what is left; it is for the alien, the orphan and the widow. Remember that you were a slave in the land of Egypt; therefore I am commanding you to do this." Similarly, Leviticus 19:33–34 states: "When an alien resides with you in your land, you shall not oppress the alien. The alien that resides with you shall be to you as the citizen among you; and you shall love him as yourself; for you were aliens in the land of Egypt." However, in Deuteronomy 20:10–18, the same charitable God offers the following instructions concerning other groups:

When you draw near to a town to fight against it, offer it terms of peace. If it accepts your terms of peace and surrenders to you, then all the people in it shall serve you at forced labor. If it does not submit to you peacefully, but makes war against you, then you shall besiege it; and when the LORD your God gives it into your hand, you shall put all its males to the sword. You may, however, take as your booty the women, the children, livestock, and everything else in the town, all its spoil. You may enjoy the spoil of your enemies, which the LORD your God has given you. Thus you shall treat all the towns that are very far from you, which are not the towns of the nations here. But as for the towns of these peoples that the LORD your God is giving you as an inheritance, you must not let anything that breathes remain alive. You shall annihilate them—the Hittites and the Amorites, the Canaanites and the Perizzites, the Hivites and the Jebusites—just as the LORD your God has commanded, so that they may not

teach you to do all the abhorrent things that they do for their Gods, and
you thus sin against the LORD your God.

During this period of their history, the People of Israel were a warrior
culture, and thus it is unsurprising that they should be instructed by their
religion to behave very differently among themselves than toward members of other groups. The same can be said for most ethnic groups, who
often linguistically restrict the term "human" to themselves (Levine and
Campbell 1972). Modern religious groups that are not ethnically based are
no different. When a Catholic priest criticized the Protestants of Geneva in
1533, he in turn was criticized for "blasphemy against God, the faith and
us, wounding our honour and calling us Jews, Turks and dogs" (Collins
1968, 87). I am not being original in pointing out these well-known facts.
However, evolutionary biology in general and multilevel selection theory
in particular may account for the facts better than any other intellectual
and scientific framework.

There is a widespread tendency to regard in-group morality as hypocritical, leading to a form of moral outrage that becomes especially intense
when applied to Judaism (e.g., Hartung 1995a, b). After all, isn't it the
ultimate in hypocrisy for a religion to simultaneously preach the Golden
Rule and instruct its members to commit genocide? This double standard
is indeed hypocritical from a perspective that envisions all people within
the same moral circle. I am being sincere when I say that this perspective
is laudable, important to work toward in the future, and possible at least
in principle to implement. However, it provides a poor theoretical foundation for understanding the nature of religions and other moral systems as
they exist today and in the past. As we have already seen, multilevel selection theory is uniquely qualified to predict both the benign nature of
within-group morality and at least three forms of human conduct that
appear immoral from various perspectives: conduct toward other groups,
the enforcement of moral rules within groups, and the self-serving viola-
tion of moral rules within groups. Multilevel selection theory accounts for
the double standard of the Hebrew Bible rather than merely reacting to
it as hypocritical. No other theoretical framework fits the well-known facts
of Judaism and other religions so well, or so I claim.

Although the double standard of the Hebrew Bible is typical of religions and ethnic groups in general, Judaism is more remarkable in other
respects. Most cultures and ethnic groups last for mere centuries before
disappearing as recognizable entities by mingling with other cultures and
ethnic groups. In contrast, Judaism has maintained its cultural identity for

thousands of years against the greatest possible odds, as the religion of a landless people dispersed among many nations. It is easy to explain the persistence of a culture that is protected by military might or geographical barriers, but something about Judaism has proved stronger than the sword or even mountain ranges and oceans. Two questions need to be asked: First, how did Jewish communities remain culturally isolated within their host nations? Second, given their cultural isolation, how did Jewish communities survive despite frequent persecution?

Cultural isolating mechanisms

Cultural and ethnic minorities are frequently isolated by the larger cultures within which they are embedded. Jewish communities have certainly experienced their share of this kind of isolation. However, many versions of Judaism also include very strong isolating mechanisms within their own structure. Once again, evolutionary biologists are not being original by pointing out this well-known fact, which is readily acknowledged by practicing Jews. Prager and Telushkin (1983) even regard it as offensive to suggest that Judaism is defined by anti-Semitism, such that the elimination of the latter would cause the former to also disappear.

In chapter 2, I described Iannaccone's (1992, 1994) prediction that strict religions become strong by isolating their members from the rest of society and by making internal cooperation the only game in town. Isolation is a design feature of the religion, not an external constraint. The history of Judaism provides many examples of this general principle. Isolating mechanisms such as circumcision, dietary restrictions, dress, language, and specific laws against sexual and social interactions are well known. Here is one example from the Synod of Frankfort in 1603: "If it is proven that any Jew has drunk wine in the house of Gentile, it shall be forbidden for any other Jew to marry his daughter, or to give him lodging, or to call him to the Torah or to allow him to perform any religious function" (Finkelstein 1924, 260).

Once again, it is important to remember that Judaism, like other major religious traditions, exists in many specific versions that vary along a spectrum from extreme separation to extreme accommodation. This spectrum has existed throughout the history of Judaism in addition to the present day, as I will describe in more detail in chapters 5 and 6. Nevertheless, the strictest and strongest versions of Judaism can accurately be described as cultural fortresses that kept outsiders out and insiders in.

The degree to which Jewish communities were isolated from their host cultures is even reflected at the level of gene frequencies. Population genetics data allow this fact to be determined with a high degree of certainty: Jewish populations from around the world are genetically more similar to each other and to the Middle Eastern population from which they were derived than to the populations among which they currently reside (Cavalli-Sforza and Carmelli 1977; Kobyliansky et al. 1982; Hammer et al. 2000). To appreciate the difference between cultural and genetic isolation, consider the Nuer, whose religion and territorial expansion at the expense of the Dinka tribe were described in chapter 2. Although some Dinka were killed during Nuer raids, many others were incorporated into the Nuer tribe: captured women as wives, men as slaves or even as substitute husbands for Nuer widows in a custom called ghost marriages (Kelly 1985). The Nuer remained distinct as a culture because assimilated Dinka abandoned their old ways and adopted Nuer ways. However, the two cultures were not genetically distinct, since fully half the Nuer were descended from ancestors who were Dinka a mere two or three generations back.[7]

So it is with most cultures around the world, whose distinctness as cultures says little or nothing about barriers to gene flow. Many factors provide genetic bridges across cultural boundaries. People who convert to a religious faith change only their beliefs, not their genes. Men have a strong incentive to acquire women from outside their own culture, by force if necessary, and to legitimize their own offspring. Adultery, prostitution, and rape also transport genes across cultural barriers. Even a weak flow of genes is sufficient to eliminate genetic differences among groups when compounded over many generations. The fact that Jewish populations around the world are genetically more similar to each other than to the populations among which they reside therefore demonstrates an extraordinary degree of isolation achieved by cultural mechanisms. Especially interesting is the fact that Judaism opposed the biological interests of the most powerful members of the community—men—by restricting their ability to import women from elsewhere. The genetic data show that these constraints were largely successful.

Judaism existed before the advent of Christianity and Islam, which were designed to grow by conversion. It has always been possible to convert to Judaism (the Hebrew Bible provides numerous examples) but only with great difficulty. In a sense, this is exactly what Iannaccone would predict for a church that wants to remain strong by forcing its new members to demonstrate their commitment. Many religious sects are hard to join. Fraternity rites and high membership costs for exclusive clubs provide examples for nonreligious groups. However, these organizations

usually seek new members, however demanding their initiation procedure. In contrast, Jewish communities almost never sought converts, even
though they would accept them. Evidently there are no examples of Jewish
missionaries or texts written to recruit outsiders to the faith. In addition,
Jewish law sometimes accorded inferior status to converts. As one example
(described in MacDonald 1994, 67), if it was necessary to make a choice
among members of the community for compensation of property, redemption of captives, or saving lives, the order was priest, Levite, Israelite,
mamzer, Nethin, convert, and freedman. Converts ranked below the offspring of illegitimate relatives (mamzerin) and individuals from foreign
ethnic groups that lived as servants among the Israelites (Nethinim). These
kinds of disincentives to become Jewish, often combined with even
stronger disincentives from anti-Semitic gentile communities, make it unsurprising that outsiders usually remained such.

Judaism thus includes many features that have caused Jewish communities to remain distinct throughout their history, even as tiny minorities subject to extreme persecution and pressure to assimilate. To survive, however,
these Jewish communities needed to acquire resources, which involved
interacting with the very people from whom they were remaining apart.

Surviving intergroup interactions

According to Darwin's basic scenario for group selection, cooperation
is a fragile flower as far as within-group interactions are concerned, but
cooperative groups robustly outcompete less cooperative groups. If Jewish
communities were exceptionally cooperative by virtue of their religion,
compared to the societies with which they interacted, this would give them
an advantage in any endeavor that requires coordinated action. Their survival amidst other nations—at least in the absence of persecution—would
be assured.

This is a sensitive issue because anti-Semitism is frequently fueled by
the accusation that Jews collaborate with each other to deprive non-Jews
of their resources. Jewish conspiracy theories reach outrageous proportions
that include blatant falsehoods. *The Protocols of the Elders of Zion*, which
supposedly outlines a program of world domination, was proven to be a
forgery in 1921, but this did not prevent millions of copies from being
printed and distributed by anti-Semites ranging from Henry Ford, to Adolf
Hitler, to modern Arabic leaders (Cohn 1966; Prager and Telushkin 1983).
Accusations of within-group cooperation and between-group exploitation
were used to justify the Holocaust in Germany and countless other efforts

to eliminate the Jewish religion through murder, forced assimilation, and displacement (Prager and Telushkin 1983; MacDonald 1994, 1998a).

I hope it is obvious that these acts are morally reprehensible, although dismayingly typical of between-group interactions in general.[8] In the aftermath of World War II, psychologists made it an urgent priority to understand why people so easily adopt the kind of us/them mentality that allows atrocities such as the Holocaust to occur. Jewish psychologists such as Henri Tajfel, himself a Holocaust survivor, were at the forefront of this movement, which became known as social identity theory (see Abrams and Hogg 1990 for a recent review and Tajfel 1981, 1–2 for an autobiographical account). The main conclusion to emerge was that us/them thinking can be triggered extremely easily in normal people. The seeds of genocide are within all of us.

Social identity theory was developed in the optimistic spirit that science can help improve the human condition, despite its often sobering conclusions. Multilevel selection theory is the perfect compliment to social identity theory and needs to be approached in the same spirit. It provides the deep evolutionary explanation for why us/them thinking is so easy to invoke in normal people. It reveals the fault lines of moral reasoning that cause people to commit unspeakable acts with a clear conscience. These are not pleasant thoughts, but they must be confronted to discover practical solutions that do, in principle, exist. One purpose of this book is to argue that cultural evolution is an ongoing process capable of discovering genuinely new solutions, even out of old parts. When it comes to evolution, the fact that something hasn't happened before is a poor argument that it can't happen in the future. Let us now return to the subject of Judaism in this constructive spirit.

There are three reasons to expect Jewish communities during the Diaspora to have achieved higher degrees of cooperation than the other social groups with whom they interacted. The first is the strictness of their religion and degree of isolation that I have already recounted. The second is the degree of genetic relatedness that often existed within Jewish communities. Kin selection has long been recognized as a powerful mechanism for the evolution of altruism (Hamilton 1964, 1975). In chapter 1, I explained that kin selection is a kind of group selection and that moral systems enable another kind of group selection that can take place among unrelated individuals.[9] It follows that social groups whose members are both genetically related and unified by a moral system should be very cooperative indeed. Jewish communities were often founded by very small numbers of individuals who themselves were highly related to each other, since uncle-niece and first cousin marriages were often encouraged. The many mechanisms employed by religions to encourage brotherhood within groups could build upon a foundation of genetic brotherhood in the case of Judaism.

The third reason to expect cooperation within Jewish communities involves ongoing group-level selection. In chapter 3, I suggested that the Protestant Reformation was a process of cultural group selection involving many social experiments, a few of which succeeded. The history of Judaism can be interpreted even more plausibly as a process of ongoing cultural and even genetic group selection, in which Jewish communities that fail to exhibit solidarity disappear, leaving the survivors to expand and create new communities. It would be extraordinary if the tragic persecution of Jewish communities over the last two thousand years did not result in a form of group-level selection.

The theoretical expectation that Jewish communities should be highly cooperative is amply confirmed by the historical record. Let us begin with the benign aspects of cooperation, as we did with Calvinism. Jewish communities throughout history have been legendary for their absence of crime, poverty, alcoholism, and other social problems. Practicing Jews are justly proud of these accomplishments, as indeed they should be. The following passage is from a book on anti-Semitism from a Jewish perspective:

> In nearly every society in which the Jews have lived for the past two thousand years, they have been better educated, more sober, more charitable with one another, committed far fewer violent crimes, and had a considerably more stable family life than their non-Jewish neighbors. These characteristics of Jewish life have been completely independent of Jews' affluence or poverty. As the noted Black economist Thomas Sowell has concluded: "Even when the Jews lived in slums, they were slums with a difference—lower alcoholism, homicide, accidental death rates than other slums, or even the city as a whole. Their children had lower truancy rates, lower juvenile delinquency rates, and (by the 1930s) higher IQ's than other children. . . . There was also more voting for congressmen by low income Jews than even by higher income Protestants or Catholics. . . . Despite a voluminous literature claiming that slums shape people's values, the Jews had their own values, and they took those values into and out of the slums." (Prager and Telushkin 1983, 46)

What religion or society wouldn't want to boast about these accomplishments? Prager and Telushkin stress again and again that the virtues of Judaism reside in the religion, not the people. Assimilated Jews quickly fall prey to the ills of the surrounding society. This interpretation is fully consistent with my own account of religion from a multilevel perspective. Groups require a strong moral system to function adaptively, and Judaism provides an exceptionally strong moral system.

Cooperation within groups is easy to admire, but the very same cooperation becomes morally ambiguous in the context of between-group interactions. We have seen that the Hebrew Bible instructed Jews behave honorably toward outsiders in some contexts but also to use their cooperation as a weapon against other groups. Did this double standard continue during the Diaspora? And what should we expect on the basis of multilevel selection theory?

One of the beauties of multilevel selection theory is that it employs the same principles at all levels of the biological hierarchy. Within-group selection by itself creates a world without morality in which individuals merely use each other to maximize their relative fitness. Group selection creates a moral world within groups but doesn't touch the world of between-group interactions, which remains exactly as instrumental as within-group interactions in the absence of group selection. Moral conduct among groups can evolve in principle, but only by extending the hierarchy to include groups of groups. This possibility is not as far-fetched as it may appear. Remember that individual organisms are already groups of groups of groups, if the emerging paradigm of major transitions is correct. Perhaps history will reveal the rudiments of moral conduct among human groups struggling to emerge against opposing forces, rather than the total absence of moral conduct among groups. Even so, we should expect far more naked exploitation among groups than within groups. Jewish groups were the victims of exploitation far more often than the perpetrators, but they too frequently adopted an instrumental attitude toward members of other groups. The double standard of Judaism continued during the Diaspora, although highly constrained by the opportunities available. Acknowledging this fact is not anti-Semitic but merely affirms the universality of us/them thinking that was established by social identity theory long ago.

Jewish history is not as simple as a displaced people struggling to survive amidst hostile neighbors. Jewish groups survived and even prospered through specific activities and relationships with different elements of their host nations. From a purely actuarial standpoint, periods of prosperity were required to balance the catastrophic declines caused by persecution.

A common pattern was for Jews to form an alliance with one gentile segment of the host nation, usually the ruling elite, to exploit another gentile segment, such as the peasantry. Far from being anti-Semitic, the ruling elite would attempt to protect the Jews from the rest of the resentful host population. This kind of relationship is illustrated by Joseph in the biblical account of the sojourn in Egypt:

> Joseph intercedes with the pharaoh on behalf of his family: "Then Joseph
> settled his father and his brothers, and gave them a possession in the
> land of Egypt, in the best of the land . . ." (Gen. 47:11). However, the ac-
> count also emphasizes Joseph's role in oppressing the Egyptians on behalf
> of the king. Joseph sells grain to the Egyptians during a famine until he
> has all of their money. He then requires the Egyptians to give their live-
> stock for food and finally their land. "The land became the Pharaoh's;
> and as for the people, he made slaves of them from one end of Egypt to
> the other" (Gen. 47:20–21). However, regarding the Israelites, the section
> continues: "Thus Israel dwelt in the land of Egypt, in the land of Goshen;
> and they gained possessions in it, and were fruitful and multiplied exceed-
> ingly." (MacDonald 1994, 114)

We do not know if this account is factual, but similar alliances with the ruling classes existed throughout Jewish history. Katz (1961, 55) describes the situation in sixteenth- through eighteenth-century Europe: "Since Jewish society was segregated religiously and socially from the other classes, its attitude toward them could be purely instrumental. . . . The non-Jew had no fear that the Jew would take a partisan stand in the struggle between the rulers and the ruled, who bore the economic yoke of the political privileges enjoyed by the rulers." The Jew's outsider status was an advantage as far as the rulers were concerned. This is a bleak picture from a broad-scale moral perspective, but it is exactly what we should expect from the largely amoral world of among-group interactions.

Many Jewish laws established an economic double standard with the same clarity that the Hebrew Bible prescribed military conduct toward other groups. The following example is from Maimonides' Mishnah Torah (book 13, The Book of Civil Laws, chap. 5:1, 93; quoted in MacDonald 1994, 148):

> It is permissible to borrow from a heathen or from an alien resident and
> to lend to him at interest. For it is written *Thou shalt not lend upon interest*
> *to thy brother* (Deut. 23:20)—To thy brother it is forbidden, but to the rest

of the world it is permissible. Indeed, it is an affirmative commandment to lend money at interest to a heathen. For it is written *Unto the heathen thou shalt lend upon interest* (Deut. 23:21).

To pick another example from more recent times, Ashkenazi Jews were not allowed to underbid other Jews for franchises or to interfere with Jewish monopolies of gentile resources, to avoid losing the "money of Israel" (Katz 1961, 61). I don't mean to imply that Jewish communities during the Diaspora invariably adopted an instrumental attitude toward members of other groups. In fact, their ability to act as corporate units may have enabled them to manage their reputations in cooperative intergroup relations more successfully than other groups, since dysfunctional groups have such poor control over their members that they are unable to maintain a reputation, even if they want to. Nevertheless, it is safe to say that Jewish communities during the Diaspora frequently engaged in economic between-group competition, and that their strong religion gave them the decisive edge that comes when more cooperative groups compete against less cooperative groups.

To summarize, multilevel selection theory provides a panoramic view of Judaism as a religion that has existed in the form of highly segregated groups which have harnessed the power of cooperation to succeed at almost everything that requires coordinated action. Cooperation includes within-group interactions that are admirable in every respect and among-group interactions that are basically competitive and therefore objectionable to the members of the other groups. Among-group interactions may exhibit rudiments of moral conduct but are dominated by exploitation on all sides. It is incorrect to say that anti-Semitism is a reaction to unfair Jewish practices because that implies that anti-Semitic groups are moral agents in their own right. Multilevel selection predicts, and history shows, that the Jews would be targets for exploitation no matter how exemplary their conduct toward other groups because groups are not moral agents unless special conditions are met. The history of Judaism during the Diaspora has been the history of growth fueled by cooperation and decline fueled by persecution in myriad forms, from the restriction of economic opportunities to forced assimilation, expulsion, and genocide.[10]

Truth in fiction

I have included Judaism in this chapter as one of three examples that establish the obviousness-in-retrospect of religious groups as adaptive units. I doubt that a byproduct theory of religion or a theory based on within-group advantage can even remotely explain the facts of Judaism as well as a theory based on group-level advantage. Multilevel selection theory also fits hand-in-glove with social identity theory, which began as an effort to understand how the human mind can be capable of atrocities such as the Holocaust. Both theories force us to confront the uncomfortable truth that us/them thinking is a part of normal human psychology. Most of us, perhaps even all of us, are capable of restricting our moral conduct to a subset of the human race and of behaving instrumentally toward outsiders. This generalization applies to all human groups and should never be used as a tool of aggression against members of a given religion such as Judaism.

I will end my discussion of Judaism in a somewhat unorthodox fashion by describing a work of fiction. Great novelists do not employ the scientific method, but they strike such a chord of truth that they are read and beloved across time and cultural boundaries. Isaac Bashevis Singer was such an author whose many novels about Jewish life in Poland before World War II and in America after the war earned him the Nobel Prize for literature in 1978. No one can accuse Singer of being an anti-Semite, yet he saw the moral world of within-group interactions and the largely amoral world of among-group interactions with a clarity and compassion that scientists and intellectuals would do well to emulate. His historical novel *The Slave* (Singer 1962) is set in seventeenth-century Poland and parallels the academic literature I have reviewed in uncanny detail.

The Slave centers on Jacob, a devout Jew who was captured and carried away by Cossacks during one of their recurrent massacres and sold as a slave to a Polish peasant. Jacob's wife and children were killed and most of his village was destroyed in the same massacre. He spends the warm months alone on a mountain tending his master's cows. For years he struggles to keep his faith and observe its many laws that set him apart from the peasants. Indeed, he has little incentive to join them because he is disgusted by their brutish ways. The strong dominate the weak and have flocks of bastard children. Offspring openly welcome the death of their own parents. During one scene of debauchery that he is forced to attend, Jacob reflects on the seeming cruelty of his own God toward the heathen:

> Soon an emissary stepped forward, a cowherd who was more fluent than
> the others, and who assured Jacob no harm was intended. They had
> merely come to invite him to drink and dance with them. The man drib-

bled, stammered, mispronounced words. His companions were already drunk and laughed and screamed wildly. They held their stomachs and rolled about on the ground. He would not be let off this time, Jacob knew.

"All right," he finally said, "I'll go with you, but I'll eat nothing."

"Jew, Jew. Come. Come. Seize him. Seize him."

A dozen hands grasped Jacob and started to tug him. He descended the hill on which the barn was located, half running, half sliding. An awful stench rose from that mob; the odors of sweat and urine mingled with the stink of something for which there is no name, as if these bodies were putrefying while still alive. Jacob was forced to hold his nose and the girls laughed until they wept. The men hee-hawed and whinnied, supported themselves on each other's shoulders, and barked like dogs. Some collapsed on the path, but their companions did not pause to assist them, but stepped over the recumbent bodies. Jacob was perplexed. How could the sons of Adam created in God's image fall to such depths? These men and women also had fathers and mothers and hearts and brains. They too possessed eyes that could see God's wonders.

Jacob was led to a clearing where the grass was already trampled and soiled with vomit. A keg of vodka three-quarters empty stood near an almost extinguished fire. Drunken musicians were performing on drums, pipes, on a ram's horn very like that blown on Rosh Hashana, on a lute strung with the guts of some animal. But those who were being entertained were too intoxicated to do more than wallow on the ground; grunting like pigs, licking the earth, babbling to rocks. Many lay stretched out like carcasses. There was a full moon in the sky, and one girl flung her arms around a tree trunk and cried bitterly. A cowherd walked over, threw branches on the fire, and nearly fell into the flames. Almost immediately a woolly looking shepherd attempted to put out the blaze by urinating on it. The girls howled, screamed, cat-called. Jacob felt himself choking. He had heard these cries many times before, but each time he was terrified by them.

"Well, now I have seen it," he said to himself. "These are those abominations which prompted God to demand the slaying of entire peoples."

As a boy, this had been one of his quarrels with the Lord. What sin had been committed by the small children of the nations Moses had been told to annihilate? But now that Jacob observed this rabble he understood that some forms of corruption can only be cleansed by fire. Thousands of years of idolatry survived in these savages. Baal, Astoreth, and Moloch stared from their bloodshot and dilated eyes. (Singer 1962, 56–57)

Nevertheless, Jacob cannot help falling in love with Wanda, his master's daughter, who walks up from the mountain every day to collect the milk from the cows. She also loves him for being endlessly more refined and learned than her own people. Finally they succumb to their passion, but Jacob's joy is mixed with agony for having committed the gravest of sins. No sooner have they embarked upon their hopeless affair when Jacob is snatched away by his own brethren, who were told of his whereabouts by a traveling circus and who arrive in a wagon with enough money to buy him back. Their altruism is unsurprising to Jacob because "to free a captive is considered a holy act" (49). His departure is so abrupt that he cannot say goodbye to Wanda, much less explain his dilemma to his rescuers. Neither does Wanda learn the truth from her own people, who lie to keep the ransom money. As far as she knows, Jacob has simply disappeared and was likely killed by some of her neighbors who resented the presence of a Jew in their midst all along.

Jacob returns to his village to find it rebounding from the massacre, as it and other Jewish communities had so often in the past. Here is how Singer describes the vitality and collective organization that caused the Jews to excel at everything that was not specifically denied to them:

> Despite the upheaval, Poland's commerce remained in the hands of the Jews. They even dealt in church decorations, although this trade was forbidden them by law. Jewish traders traveled to Prussia, Bohemia, Austria, and Italy, importing into the country silk, velvet, wine, coffee, spices, jewelry, weapons, and exporting salt, oil, flax, butter, eggs, rye, wheat, barley, honey, hides. Neither the aristocracy nor the peasantry had any real knowledge of business. The Polish guilds continued to protect themselves through every form of privilege, but nevertheless their products were more expensive than those of the Jews and often inferior in workmanship. Nearly every manor harbored Jewish craftsmen, and, although the king had forbidden Jews to be apothecaries, the people had confidence in no others. Jewish doctors were sent for, sometimes from abroad. The priests, particularly the Jesuits, harangued against infidel medicine from their pulpits, published pamphlets on the subject, petitioned the Sejm and the governors to disqualify the Jews from medical practice, but no sooner did one of the clergy fall ill, than he called in a Jew to attend him. (126–27)

Jacob is immediately given a place in the community, but he cannot find happiness. So many of his brethren merely observe the rituals of their faith while violating its spirit to enrich themselves. Although penniless, as a scholar he is an attractive marriage prospect, but none of the matches arranged for him have the passion or even the spirituality of his love for Wanda. At great risk and even greater spiritual agony for "returning to Egypt," Jacob returns for Wanda and they escape through the mountains to a newly founded Jewish community where they are both strangers. Even though Wanda wants to convert and is being instructed by Jacob, Jewish law would never permit it and she couldn't possibly learn the language and customs well enough to masquerade as one who is already a Jew. The only way they can remain together is for her to pretend to be deaf and dumb.

The new Jewish community has formed on the estate of an inept Catholic nobleman who originally invites only a few families to manage his affairs. This arrangement is typical; in fact, the reason that Jacob can speak fluent Polish is because his mother's side of the family has managed the affairs of the Polish nobility for generations. In no time the tiny Jewish settlement mushrooms into a town that grows in front of the helplessly protesting nobleman's eyes. His Jewish manager not only unscrupulously exploits him and the peasants at every opportunity, but also does the same to his own people while piously observing the ritual aspects of his faith. As a genuinely pious man, Jacob is thrust against his will into the position of manager, which threatens even more to expose his horrible charade.

I will not reveal the dramatic end of this wonderful story. Singer's genius in all of his novels was to observe human affairs with a clinical eye without losing his compassion for humanity. In *The Slave*, he turns his clinical eye toward the same within- and between-group interactions that I have recounted from the scholarly literature and tried to explain from a multilevel evolutionary perspective. A work of fiction does not confirm a scientific theory, but Singer provides a perfect summary of my own thesis when he says through the character of Jacob: "But now he at least understood his religion: its essence was the relation between man and his fellows" (247).

Example 3: The Early Christian Church

The Jews were one of many cultures caught in the embrace of the Roman Empire two thousand years ago. Christianity began as a tiny Jewish sect and grew into a mighty religious tree of which Calvinism is just one branch. Can the original Christian Church be analyzed in the same functional terms as the Judaism that came before and Calvinism that came after? A remarkable book by Rodney Stark (1996) suggests that it can. The fact that I criticized Stark's rational choice theory of religion in chapter 2 and rely upon his work here to support my own thesis returns us to the problem of multiple perspectives that so vastly complicates the scientific process.

Whatever I think of Stark's theory, I can only admire his skill as a scientist working with fragmentary data. Consider the seemingly miraculous growth of the early Christian Church, from a tiny sect of approximately 1,000 members in the year 40 to such a large fraction of the Roman Empire in year 350 that the emperor Constantine himself converted. Stark shows that the main miracle at work here is the miracle of exponential growth. A constant growth rate of only 40 percent per decade is sufficient to explain the increase, which is commensurate with some of the fastest growing modern religions (e.g., Mormonism). Thus, there is no need to invoke extraordinary processes such as mass conversions to explain the growth of the early Christian Church. The normal processes that can be observed in modern religious movements are probably sufficient.

One of these normal processes is growth through pre-existing social networks. A person is far more likely to convert to a new faith if a friend, relative, or other social partner has previously converted. In this very local sense, conversion can preserve social ties rather than breaking them, no matter how divorced from society the sect may be at a larger scale. Stark uses this principle, which he developed himself based on observations of the "Moonies" in the 1960s, to argue that the early Christian Church was much like the Reform Judaism that began in Europe and America in the nineteenth century. In both cases, large numbers of Jews found it difficult to conform to the strict laws of Judaism and had powerful incentives to create social ties with gentiles. Yet, they did not want to abandon the great benefits associated with the Jewish religion. Reform Judaism explicitly attempts to separate theology from ethnicity and even proclaims itself a religious community open to new members, rather than a nation that one is primarily born into (Stark 1996, 53–55). According to Stark, early Christianity had exactly the same appeal for Jews during the Roman period and also for gentiles who admired the Jewish faith (numerous enough to be given the name of "God-fearers" at the time) but who could not become full-fledged Jews.

This line of reasoning is controversial because the mission to the Jews is supposed to have failed early in the history of the Christian church, leaving it primarily a gentile movement. Fortunately, Stark goes beyond proposing a controversial new theory and proceeds to test it with a quantitative analysis of twenty-two Roman cities. Using the date at which Christian churches first appeared as the dependent variable, he explains an impressive 67 percent of the variance with the independent variables city size, distance along trade routes between Jerusalem and Rome, and the presence of a synagogue around the year 100 as a measure of Jewish influence within each city. Christianization correlates positively with Jewish influence and negatively with Roman influence, as expected from Stark's hypothesis. He even resolves a lesser controversy among religious historians by showing that Gnosticism (another religious movement of the period) is probably an offshoot of Christianity rather than a parallel offshoot of Judaism.

Stark makes sense of early Christianity the way that the theory of glaciation made sense of the valley of Cwm Idwal for Darwin. What before was a jumble of uninterpretable facts suddenly becomes organized by the relationships made reasonable by the theory. Even isolated tidbits of information, such as the location of Christian churches in the Jewish sections of cities and rental receipts between Jewish landlords and Christian tenants, add weight to the theory by falling into place, like disparate clues coming together in a detective story. Stark combines his scientific skills with a genuine respect for the scholars from many disciplines whose painstaking work provides the information upon which he depends. Little wonder, then, that his book has earned the same kind of praise and respect in the normally contentious field of religious studies that Lansing's book on the water temple system of Bali earned among anthropologists.

Why am I lavishing such praise on the very same person whose work I criticized in chapter 2? Remember that my criticism of the rational choice literature in chapter 2 was highly specific: it seems to reject functionalism when in fact it does nothing of the sort. The same literature is highly praiseworthy in other respects. Although Stark's formal theory is strongly committed to a byproduct view of religion, the larger corpus of his work is not. Let us see how Stark paints a vivid portrait of the early Christian Church as an organismic unit adapted to survive and reproduce in the chaotic environment of the Roman empire.

Chaotic it was. Most of us imagine the typical Roman city as like the movie set for Ben Hur, but in fact they were incredibly crowded, filthy, and prone to catastrophic disaster. Stark lists the disasters that befell the city of Antioch:

> During the course of about six hundred years of intermittent Roman rule,
> Antioch was taken by unfriendly forces eleven times and was plundered

and sacked on five of these occasions. The city was also put to siege, but did not fall, two other times. In addition, Antioch burned entirely or in large part four times, three times by accident and once when the Persians carefully burned the city to the ground after picking it clean of valuables and taking the surviving population into captivity. Because the temples and many public buildings were built of stone, it is easy to forget that Greco-Roman cities consisted primarily of wood-frame buildings, plastered over, that were highly flammable and tightly packed together. Severe fires were frequent, and there was no pumping equipment with which to fight them. Besides the four huge conflagrations noted above, there were many large fires set during several of the six major periods of rioting that racked the city. By a major riot I mean one resulting in substantial damage and death, as distinct from the city's frequent riots in which only a few were killed.

Antioch probably suffered from literally hundreds of significant earthquakes during these six centuries, but eight were so severe that nearly everything was destroyed and huge numbers died. Two other quakes may have been nearly as serious. At least three killer epidemics struck the city—with mortality rates probably running above 25 percent in each. Finally, there were at least five really serious famines. That comes to forty-one natural and social catastrophes, or an average of one every fifteen years. (159)

To make matters worse, the people that filled Roman cities belonged to a diversity of ethnic groups that hated each other. Not only did walls surround the city of Antioch to keep out unfriendly forces, but they also existed *within* the city to divide ethnic factions, which included Macedonians, Cretans, Cypriotes, Argives, Herakleidae, Athenians, Syrians, and Jews (157). As expected on the basis of the previous section of this chapter, the Jewish segment of the population expanded markedly as the city grew (Meeks and Wilken 1978). I cannot improve on Stark's final summary of the urban social environment that formed the background for early Christianity:

Any accurate portrait of Antioch in New Testament times must depict a city filled with misery, danger, fear, despair, and hatred. A city where the average family lived a squalid life in filthy and cramped quarters, where at least half of the children died at birth or during infancy, and where most of the children who lived lost at least one parent before reaching maturity. A city filled with hatred and fear rooted in intense ethnic antago-

nisms and exacerbated by a constant stream of strangers. A city so lacking
in stable networks of attachments that petty incidents could prompt mob
violence. A city where crime flourished and all the streets were dangerous
at night. And, perhaps above all, a city repeatedly smashed by cataclysmic
catastrophes: where a resident could expect literally to be homeless from
time to time, providing that he or she was among the survivors. (160–61)

Against this background, early Christian society must have looked very
good indeed. For any coherent culture to survive amidst such chaos, it
must possess some kind of an isolating mechanism. The analogy to a biological cell in instructive. Cell membranes allow wonderfully complicated
self-sustaining processes to take place inside the cell amidst a larger outside world of chaos. The organismic concept of groups encourages us to
look for something similar—a culturally defined membrane that allows
highly organized self-sustaining social interactions to take place within the
group amidst a larger world of chaos. We have seen that many forms of
Judaism possess a cultural membrane that is difficult to leave or enter.
They persist to this day because their social physiology enables them to
survive and reproduce so well in their larger environment. Perhaps the
most radical innovation of the early Christian Church was to provide a
membrane and a social physiology comparable to Judaism for anyone
who wanted to join, regardless of their ethnicity. This combination of per-
meability with respect to membership and impermeability with respect
to interactions is a remarkable piece of social engineering. Anyone could
become a Christian, but those who did were expected to overhaul their
behaviors under the direction of a single God in a close-knit community
that could easily enforce the new norms.

The behaviors prescribed by the early Christian Church were far more
adaptive than those practiced in the surrounding Roman society. We commonly assume that the sex drive and the natural urge to have children will
automatically result in babies. No one has to tell a population to grow; it
simply does grow as long as resources are available. As strange as it may
seem, Roman culture developed in a way that became hostile to biological
reproduction, despite the availability of resources. Part of the problem was
extreme male domination and a form of status-striving that made marriage
and families unattractive prospects for males. Female infanticide was so
common that Russell (1958) estimated a sex ratio of 131 males per 100
females in Rome and 140 males per 100 females in Italy, Asia Minor, and
North Africa. Preference for sons can be adaptive at the scale of a single
patriarchal lineage competing with other lineages, but it can have disas-

trous consequences at the scale of the whole society.[11] Of course, Romans had a sex drive like everyone else, but they found ways to satisfy it in ways that did not lead to reproduction, such as homosexuality and nonreproductive heterosexual practices. Women who did conceive often did so in circumstances that caused them to get an abortion, itself a life-threatening operation. We are familiar with all of these practices in modern life but seldom think about their consequences for population growth. If we did, we might see some of them as solutions to the modern problem of overpopulation. For the Romans, these practices led to a crisis of underpopulation. Julius Caesar attempted to stimulate reproduction by awarding land to fathers of three or more children and considered legislation outlawing celibacy. Similar policies were attempted by subsequent emperors but to no avail. According to Boak (1955; discussed in Stark, 116) "[policies with] the aim of encouraging families to rear at least three children were pathetically impotent." By the start of the Christian era, the Roman population had started to decline, even during the good times between plagues, and required a constant influx of "barbarian" settlers to maintain itself.

In contrast, the Christian religion, like the Jewish religion from which it was derived, expected marriage, abundant children, and fidelity in both sexes while outlawing abortion, infanticide, and nonreproductive sexual practices. When stated as a religious imperative and enforced by the social control mechanisms that come naturally to small encapsulated groups, Christianity succeeded at changing reproductive behavior as Roman law never could. Christian women raised more babies than their pagan counterparts.

Whatever modern women may think of modern Christian religions, Roman women found the early Christian Church highly attractive, not only for its rules governing reproduction but also for the status and power they could achieve. In this respect, early Christianity departed from Judaism in addition to paganism. The letters of Paul refer to women deaconesses and other women who did much more than make babies, such as Prisca, whom Paul acknowledges for having "risked her neck" on his behalf (Stark 1996, 108). In later centuries the Christian Church became more sexist and even changed the wording of the New Testament, referring to women deaconesses as servants or wives of deacons. In reality, the original sources are full of tales of how women of all ranks and especially the upper ranks of society flocked to join the early Christian Church. Thus, not only did Christian women have more offspring than pagan women, but Christian society included a higher proportion of women than pagan

society. Part of the rise of Christianity can be attributed to biological reproduction, pure and simple.

Differential survival also played a key role. Two plagues swept through
the Roman Empire during the first three centuries of Christian history.
Stark estimates that between a quarter and a third of the Empire's population died in each case. No one at the time knew about germs, so it might
seem that a plague would fell pagan and Christian alike. However, it turns
out that simple nursing practices can make the difference between life and
death for a disease such as smallpox or measles, which are suspected to
have been the agents of the two Roman plagues. Simple provision of food
and water for those too sick to cope for themselves can reduce mortality
by two-thirds or even more by modern estimates (89). Pagans and Christians alike may have caught the disease, but they differed vastly in their
response to it, as Dionysius described in a tribute following the second
epidemic around year 260:

> Most of our brother Christians showed unbounded love and loyalty,
> never sparing themselves and thinking only of one another. Heedless of
> danger, they took charge of the sick, attending to their every need and
> ministering to them in Christ, and with them departed this life serenely
> happy; for they were infected by others with the disease, drawing upon
> themselves the sickness of their neighbors and cheerfully accepting their
> pains. Many, in nursing and curing others, transferred their death to them-
> selves and died in their stead. . . . The best of our brothers lost their lives
> in this manner, a number of presbyters, deacons, and laymen winning
> high commendation so that death in this form, the result of great piety
> and strong faith, seems in every way the equal of martyrdom. (82)

Dionysius continued by describing how the non-Christians responded to
the plague:

> The heathen behaved in the very opposite way. At the first onset of the
> disease, they pushed the sufferers away and fled from their dearest, throw-
> ing them into the roads before they were dead and treated unburied
> corpses as dirt, hoping thereby to avert the spread and contagion of the fa-
> tal disease; but do what they might, they found it difficult to escape.

This account by itself cannot be taken as fact, but the weight of evidence
supports it. As one example, Galen, the great Roman physician whose

views on the heart and circulatory system remained dogma for centuries until challenged by William Harvey, responded to the plague by retiring to his villa in Asia Minor until it was over.

Altruism and social dilemmas have been discussed so much by biologists and social scientists that the ideas have lost much of their force. For altruism we trot out weary examples of birds calling at the sight of predators or soldiers falling on grenades. The most recent theories place more emphasis on punishment and social control as solutions to social dilemmas than altruism, which seems eternally vulnerable to the free-rider problem.[12] A plague forces us to confront these issues with less glibness. Imagine that you lived in Roman times and that people were dying horribly all around you. You don't know about germs but you do know about contagion, which means that even the simplest act of kindness, such as helping a plague victim drink water, will substantially increase your own chance of dying a horrible death. Knowing all of this, what would it take for you to care for your own child? Your grandparent? Your neighbor? A total stranger? How about properly disposing of the dead, who are beyond help but whose festering bodies are spreading the disease? Can you imagine doing it because it is required by law and you will be sent to jail if you don't? Or because women will regard you as sexy and your status will go up in the eyes of your peers? Something extraordinary was required to solve this social dilemma: not a divine miracle but a miracle of psychological and social engineering that Roman society lacked and Christian society provided.

As for the plagues, so for the rest of life. Christian society provided "a miniature welfare state in an empire which for the most part lacked social services" (Johnson 1976, 75; quoted in Stark 1996, 84). Even the emperor Julian acknowledged this fact in a letter to a pagan priest: "The impious Galileans support not only their poor, but ours as well, everyone can see that our people lack aid from us" (84). Julian saw the problem and tried to institute pagan charities to rival Christian charities, but the social dilemmas implied by the word "charity" are not solved so easily. Stark (87) invites us to read the following familiar passage from Matthew (25:35–40) and try to imagine how it must have appeared when it was a truly new morality:

> For I was hungry and you gave me food, I was thirsty and you gave me drink, I was a stranger and you welcomed me, I was naked and you clothed me, I was sick and you visited me, I was in prison and you came to me. . . . Truly I tell you, just as you did it to one of the least of these who are members of my family, you did it to me.

According to Stark, written accounts from the period affirm that the Christians put these words into action. A study of tombstone inscriptions also suggests that Christians had a longer life expectancy than pagans (Burn 1953). In a world where survival depended on helping, Christians survived longer than pagans in addition to having more offspring. These basic biological parameters contributed significantly to the growth of the early Christian Church, in addition to high conversion rates. What enabled Christians to practice this new morality? Stark emphasizes beliefs and social organization, just as I did in my analysis of Calvinism. Of course, the central fact and interpretation of Christ's death makes altruism the defining feature of being Christian.

Throughout his voluminous writings, Stark interprets supernatural agents as imaginary providers of resources that are scarce or impossible to obtain in the real world. The gods don't actually provide the resources; belief in their existence is merely a byproduct of a general human propensity to explain their world (propositions 1, 2, and 3 in table 2.1). The ultimate scarce resource, denied to rich and poor alike, is life after death. We might therefore predict all religions to include the promise of a glorious world that awaits us beyond the veil. Of course, even a passing knowledge of world religions suffices to reject this hypothesis. The Nuer thought little about an afterlife, despite their fear of death, and no one could look forward to the afterlife imagined by the Greeks and Romans. Even Judaism, the religion from which Christianity is derived, focuses more on establishing the nation of Israel on earth than on what happens after death. Belief in a wonderful heaven must therefore be explained by a different set of principles than a general desire to explain the world and to obtain scarce resources. In his analysis of Christianity, Stark (1996, 80–81) emphasizes the *secular utility* of belief in the afterlife, as *an adaptation to a particular environment,* quoting with approval the following passage from McNeill (1976, 108):

> Another advantage Christians enjoyed over pagans was that the teachings of their faith made life meaningful even amid sudden and surprising death. . . . Even a shattered remnant of survivors who had somehow made it through war or pestilence or both could find warm, immediate and healing consolation in the vision of a heavenly existence for those missing relatives and friends. . . . Christianity was, therefore, a system of thought and feeling thoroughly adapted to a time of troubles in which hardship, disease, and violent death commonly prevailed.

As we have seen, this system of thought not only relieved psychological stress but also motivated the death-embracing altruism that actually increased the survival of the group. Along with Durkheim, I predict that most enduring religions survive on the basis of their secular utility. Their design features include belief systems that, no matter how otherworldly, have the effect of motivating adaptive behaviors in this world. Many motivating belief systems are possible, and not all of them must include a glorious afterlife or even supernatural agents (Durkheim defined religion in terms of the sacred and the profane, not supernatural agents). When belief in a glorious afterlife is lacking, there will be other motivating elements to take its place (e.g., establishing the nation of Israel). What will *not* be observed, or rather seldom observed, are major beliefs that have no function other than to satisfy the human urge to explain, or that actually handicap the believer by motivating dysfunctional behaviors. Stark appreciates the secular utility of Christian belief but does not generalize this insight to religion in general. If he did, his formal theory of religion would resemble functionalism more than it does.

We have seen that religion cannot survive on belief alone. A system of social coordination and control is also necessary to direct action and exclude the inevitable free-riders who are indifferent to belief. The Christian emphasis on charity and forgiveness often gives the appearance of unguarded and indiscriminate altruism. Did not the early Christians nurse and support the Roman sick and poor, in addition to their own? A closer look reveals a far more sophisticated social physiology than indiscriminate altruism, just as we saw in the cases of Calvinism and Judaism. At least three categories of people can be easily distinguished: brethren in good standing, brethren in poor standing, and outsiders. Brethren in good standing received the benefits of altruism that they were also expected to give. Brethren in poor standing were subjected to an escalated series of punishments ending in exclusion. It was the apostle Paul who said "Cast out the wicked from among you." The early Christians did indeed extend a charitable hand to outsiders, in part to bring them into the church, but not to the degree that charity was practiced within the church. If there was truly no distinction between conduct toward insiders and outsiders, the first tiny Christian communities would have evaporated in no time. In addition, joining the church involved expensive commitments that can be shown theoretically to weed out free-riders (Iannaccone 1992, 1994).

One carefully calibrated policy toward outsiders concerned intermarriage. In contrast to Judaism and many other religions, the early Christian Church did not prohibit and may have even encouraged marriages between Christians and pagans. This makes adaptive sense because pagan spouses were far more likely to convert to Christianity than the reverse. Many pagan men and even entire households of servants and slaves became Christian following marriage to a Christian woman.

Perhaps I have said enough to convince the reader that the early Christian church stands alongside Calvinism, the water temple system of Bali, and Judaism as a functionally adaptive religion. The only mystery is how Stark can be so lucid about the secular utility of the early church while advancing a formal theory that for all the world makes religion appear as a functionless byproduct of the urge to explain and relegates functionalism to the rubbish heap of history.

How to Avoid Bias

I could add to my list of functionally adaptive religions. Confucianism and Islam in both early and late manifestations would be good candidates. However, arguing from examples is vulnerable to the problem of bias. Just as any political opinion can be supported with the selective use of statistics, perhaps I am picking and choosing among thousands of religions for the few that support my position. Someone else with a different view of religion might be able to do the same. What is to prevent the debate from degenerating into an intellectual food fight?[13]

Bias is a familiar problem in science with a reliable solution: random sampling. If we compare hypotheses with a sample of religions that are not preselected with respect to the hypotheses, we can let the chips fall where they may. When Elliott Sober and I wanted to say something general about group selection and human evolution, we chose a sample of twenty-five cultures at random from the Human Relations Area File, an anthropological data base designed for cross-cultural comparison (Sober and Wilson 1998, chap. 5). Only a tiny amount of information was gathered for each culture (again without respect to the hypotheses being tested), but collectively the sample was powerful because it provided an unbiased look at the vast diversity of cultures around the world. We knew that we could throw out the entire sample and that the same strong patterns would emerge with any other random sample of twenty-five cultures chosen from the same data base.

Nothing comparable to the Human Relations Area File exists for religion, but there is an authoritative sixteen-volume encyclopedia of religion (Eliade 1987). To compile a random sample, I first wrote a computer program to select volumes at random and page numbers at random within each volume. The volume and page chosen by the computer sent me to an entry in the encyclopedia that may or may not refer to a specific religion. Many entries concern more general subjects such as "myth" or "polytheism" that discuss many religions. I therefore paged forward from the original page number until I encountered an entry that could be associated with a single religious system. This might be the name of a person who founded a new religious movement (e.g., Eisai, founder of the Rinzai shool of Zen Buddhism in Japan during the twelfth century) or the name of the movement itself (e.g., the Cao Dai cult that originated in Vietnam during the twentieth century). I decided to exclude tribal religions (thus, "Nuer" would have been passed over) and included only religions with a known historical starting date.[14] Minor religious movements within a larger religious tradition were included, since the larger traditions themselves started out as minor movements.

The twenty-five religious systems chosen in this manner are shown in table 4.1. The selection procedure is not completely unbiased, since some judgment calls were required to decide if an entry qualified as a religion. The major religious traditions may not be equally represented because some fractionate into separate movements more than others (e.g., Protestantism vs. Catholicism). More sophisticated random sampling procedures exist, but I decided to simply throw darts blindfolded at the dart board of religious systems. As it happened, all major religious traditions and most regions of the world are represented. Many of the religious systems were completely unknown to me prior to their selection.

Needless to say, the encyclopedia entry for each religion does not include sufficient information to evaluate the major issues that I have discussed at length for my selected examples. I am therefore in the process of gathering and reviewing the information that exists for each religious system, with six completed as of this writing. Even though this is work in progress, it is important to show that disciplined results are possible and that the basic functionality of religion does not need to be debated endlessly into the future, as it has in the past. Remember that this book is about science in motion, describing a game in progress rather than merely reporting the score. Another reason to discuss the survey at an early stage is to invite readers with appropriate expertise to join the game. Scholarship is, or should be, a communal effort, and there are many scholars of religion more highly qualified than I to evaluate these particular religious systems. The simple act of providing a *random* sample

---

Table 4.1 Random Sample of Twenty-five Religious Systems. Volume and page
numbers refer to entries in the *Encyclopedia of Religions* (Eliade 1987).

1. Indus Valley Religion, India, 2500 B.C.E. (7:215)
2. Eisai, founder of Rinzai school of Zen Buddhism, Japan, 1141–1215 (5:72)
3. Maurice, F. D., founder of Christian Socialism, 1805–72 (9:287)
4. Nagarjuna, Buddhist, South India, ca. 150–250 (10:290)
5. Mahavira, Jainism (9:128)
6. Cult of Saints (4:172)
7. Pelagianism, fifth-century Rome (11:230)
8. Maranke, J., Christian Apostolic Church, Africa, 1912–63 (9:188)
9. Chen-Yen, Buddhism, ca. 230 (3:230)
10. Allen, R., founder of Mother Bethel A.M.E. Church, U.S.A., 1760–1831 (1:211)
11. Lahori, founder of Lahori branch of Ahmadiyah movement, 1874–1951 (8:423)
12. Pietism, Christianity, 1675 (11:324)
13. Mbona, African cult (9:303)
14. Young, B., Mormonism, 1801–77 (15:539)
15. Cao Dai cult, Vietnam (8:451)
16. Nahman of Bratslav, Hasidic master, 1772–1810 (10:297)
17. Chinul, founder of Chogye School of Korean Son, 1158–1210 (3:333)
18. Atisa, founder of Buddhist path literature, 982–1054 (1:492)
19. Renyo, Honganji movement, Japan (12:335)
20. Neo-Orthodoxy, Protestant movement, 1940s (10:360)
21. Gokalp, Z., Turkish nationalism, 1876–1924 (6:66)
22. Theosophical Society, 1875 (14:464)
23. Dalai Lama, Tibetan Buddhism (4:200)
24. Mawdudi, S. A., Islamic revivalist movement, 1903–79 (9:291)
25. Agudat Yisra'el, Orthodox Jewish movement (1:149)

---

means that the information that accumulates for these particular religious systems will say something very important about religion in general. The entire sample could be discarded, another twenty-five darts could be thrown randomly at the dart board, and the same basic results would emerge.

Without attempting to prejudge the results, I think that the basic functionality of religion will be amply confirmed. Durkheim was right. Something as elaborate—as time-, energy-, and thought-consuming—as religion would not exist if it didn't have secular utility. Religions exist primarily for people to achieve together what they cannot achieve alone. The mechanisms that enable religious groups to function as adaptive units include the very beliefs and practices that make religion appear enigmatic to so many people who stand outside of them.

SEEING THE OBVIOUS

I began this chapter with Darwin's account of how he failed to see the evidence for glaciers that lay all around him, for lack of the right theory. By chance, the authors that I drew upon for my three selected examples of religious systems represent a number of major intellectual traditions, including Marxism, cultural relativism, evolutionary biology, and rational choice theory. These traditions are almost never related to each other in a constructive fashion; more often, their adherents speak of each other as evil (or even worse, irrational) aliens from other planets. Different images of religion also seem to emerge from each tradition; compare Stark's theory with Marx's famous statement that religion is the opium of the masses. All of these traditions confront the same world but see different worlds based on their ideas that organize perception. In this sense they really do come from different planets. But there is only one world, and with the appropriate set of ideas we can see the functional nature of religious groups as clearly as Darwin eventually saw the evidence for glaciers in the valley of Cwm Idwal.

CHAPTER 5

THE SECULAR UTILITY OF RELIGION
The Modern Literature

Truth, Karl Deutsch observed, lies at the confluence of independent
streams of evidence. The prudent social scientist, like the wise investor,
must rely on diversification to magnify the strengths, and to offset the
weaknesses, of any single instrument.

—Putnam 1992, 12

This advice is valid for all scientific inquiry, not just the social sciences.
However, it does not fit the heroic image of a contest between rival theories settled by a single decisive experiment. Darwin was roundly criticized
by many of his peers for using diverse streams of evidence to support his
theory of evolution (Hull 1973).

In previous chapters I have tried to show that traditional religious
scholarship is comparable to the natural history information available to
Darwin and that it provides ample evidence for religious groups as adaptive units. Precise measurements and statistical comparisons sharpen perception, but they are not always required to see what is already plain to
our senses, once aided by the right theory. Of course, as prudent scientists
we should examine many streams of evidence, including research on religion reported in periodicals such as the *Journal for the Scientific Study of
Religion, Review of Religious Research,* and *Annual Review of the Social Sciences
of Religion.* If the methods of science sharpen perception, then modern
research should provide even better evidence for the picture that has already emerged on the basis of traditional scholarship.

In this chapter I will review the streams of evidence that run through
the modern scientific literature on religion. In the process, I will show
how evolutionary thinking can provide a new framework for the study of

religion that goes beyond rational choice theory and other social scientific approaches.

SOLVING THE FUNDAMENTAL PROBLEM OF SOCIAL LIFE

If religious groups function as adaptive units by preventing cheating and coordinating behavior, then members of religious groups should prosper more than isolated individuals or members of less adaptively organized groups. The word "prosper" should have roughly the same meaning for religious and nonreligious folk. As I mentioned in chapter 1, to a first approximation we don't need to puzzle over the concept of fitness any more than Darwin needed to puzzle over the usefulness of a thick beak for cracking hard nuts. All farmers in Bali want an abundant rice harvest, and all people during Roman times wanted to avoid dying from plagues. Matthew 25:35–40 expresses gratitude for the basic needs of food, drink, clothing, health, and freedom from imprisonment. Religions may change some aspects of what people want, but it is built upon a foundation of providing what all people want through the coordinated action of groups.

This is a risky prediction because it can easily prove to be wrong. There is no shortage of people who regard religion as primarily dysfunctional in one way or another. Perhaps religion is a cultural parasite that deceives people into forsaking what they want in the real world for the empty promise of everlasting life. Perhaps religion is the economic mind spinning its wheels to get resources that cannot be had at the expense of resources that can be had. These theories make the opposite prediction, that religion impoverishes its members, giving only vague psychic benefits in return.

It is easy to see why religion so often appears dysfunctional, because its costs are so conspicuous. Religious folk are expected to give their time, their money, their identity, and even their lives when necessary. They are expected to forsake opportunities and adopt beliefs and practices that appear inexplicable to outsiders. These costs, however great, do not prove that religion decreases the prosperity of its members. Altruism involves sacrificing for the benefit of others. Since the benefits of the costly behaviors demanded by religion remain largely within the church, the net effect can easily increase the prosperity of the average member. As Darwin understood, groups of altruists beat groups of nonaltruists, even when judged by purely secular criteria.

I have already reviewed historical examples of religious groups that prosper in secular terms. Does modern religious research reach the same conclusion, providing independent streams of evidence? A good place to begin is with Iannaccone (1992, 1994), whose theory of religious commitment was discussed in chapter 2. He showed that strict churches can be strong by weeding out free-riders and by making religious involvement the only option for its members. According to Iannaccone, the reason a strict churches such as the Latter Day Saints is growing while a lax church such as the Episcopalians is declining in membership is because the former provides powerful benefits that more than offset its costs. It is literally a better bargain as far as net benefit is concerned.

This theory and its supporting evidence seem to reach the same conclusion as our historical examples. However, it is worth taking a closer look to reveal not only the strengths but also some of the weaknesses of modern religious research. First, can something as fuzzy-sounding as "strictness" really be measured with precision? The answer is yes. Hoge and Roozen (1979, E-4; discussed in Iannaccone 1994, 1190) had twenty-one experts (church historians, sociologists of religion, denominational leaders, and seminary educators) rate sixteen major Protestant denominations on a seven-point scale using the following operational definition of strictness: "Does the denomination emphasize maintaining a separate and distinctive life style or morality in personal and family life, in such areas as dress, diet, drinking, entertainment, uses of time, marriage, sex, child rearing, and the like? Or does it affirm the current American mainline life style in these respects?" Fifteen years later, Iannaccone replicated the survey with sixteen new experts. Not only were the denominations ranked as one would expect (with Episcopalians and Jehovah's Witnesses at the lax and strict extremes, respectively), but the correlation between the two surveys, conducted fifteen years apart, was an astounding 0.99. In addition, the rankings correlated strongly with more quantifiable measures of commitment such as church attendance. There is no doubt that modern churches can be ranked with respect to strictness, setting the stage for a quantitative test of Iannaccone's theory.

Sociologists of religion commonly break the strictness continuum into four categories: (1) liberal mainline, (2) moderate mainline, (3) conservatives and evangelicals, and (4) fundamentalists, pentecostals, and sects (Iannaccone 1992, 1994). Scores of surveys have generated reams of data on household income, level of education, church attendance, fraction of income contributed, and so on. A number of patterns emerge strongly enough to qualify as general rules. First, household income and level of education decline with increasing strictness. Members of liberal churches tend to be wealthier and better educated than members of sects. Second, the amount of money contributed to one's church increases directly with strictness. Members of sects not only contribute a higher proportion of their income than members of liberal mainline churches, but they actually contribute a higher absolute dollar amount, despite being considerably less wealthy. When financial cost is combined with the time and lost opportunity costs, there can be no doubt whatsoever that strict churches impose a burden on their members.

Superficially, these facts seem to support dysfunctional or parasitic theories of religion. Perhaps members of strict churches are less wealthy and educated because they are being bled by their churches. To support a functionalist theory of religion, we must show that strict churches provide benefits that more than offset their costs. It is here that we encounter a problem. Even though religious researchers do an admirable job of measuring strictness and the costs of religious participation, they are much less successful at measuring the benefits. In fact, costs and benefits are rarely even measured in the same currency, which makes the calculation of net benefits impossible. Here is a key passage from Iannaccone (1994, 1183), discussing the social benefits of religion that can be produced only by solving the free-rider problem:

> Religion is a social phenomenon, born and nurtured among groups of people. In principle, perhaps religion can be purely private, but in practice it appears to be much more compelling and attractive when experienced in groups. In the austere but precise language of economics, religion is a "commodity" that people produce *collectively*. My religious satisfaction thus depends on both my "inputs" and on those of others. The pleasure and edification that I derive from Sunday service does not depend solely on what I bring to the service (through my presence, attentiveness, public singing, etc.); it also depends on how many others attend, how warmly they greet me, how well they sing or recite (in English, Latin, Hebrew, Arabic, etc.), how enthusiastically they read and pray, and how deep their commitments are. The collective side of religion encompasses numerous group activities such as listening to sermons, scriptural studies, testimonial meetings, liturgies, worship, hymn singing, and sacramental acts. However, it also extends to religious belief and religious experiences—particularly the most dramatic experiences such as speaking in tongues, miraculous healings, prophetic utterances, and ecstatic trances—all of which are more sustainable and satisfying when experienced collectively.

I must confess my disappointment upon reading this passage for the first time. As a biologist, I thought that solving the free-rider problem and producing collective benefits meant warning others about predators at personal risk, cooperatively hunting and sharing food, or curtailing reproduction to conserve resources. In the context of religion I thought it meant nursing the sick, lending money without interest, sharing irrigation water with your downstream neighbor, or contributing money to buy your brethren back from slavery. All of these benefits improve the welfare of others in ways that can be measured in the same hard currency as the individual costs—time, money, opportunity, health, and ultimately survival and reproduction. Yet, Iannaccone's list is nearly devoid of material benefits, consisting instead of vague psychic pleasures, such as speaking in tongues. It is not that he denies material benefits, which are mentioned now and then throughout his papers; he merely doesn't distinguish them from psychic benefits and doesn't attach much importance to them, as the above cited passage shows. Thus, despite his ability to accurately measure strictness and the material costs of religious commitment, he does not measure benefits with the same accuracy or even in the same currency and therefore can say nothing about net benefits. His theory remains unproven.

REALITY CHECK: MATERIAL BENEFITS PROVIDED BY A MODERN CHURCH

Are the benefits of religion really so wimpy? A recent study of a Korean Christian Church based in Houston, Texas, provides a needed antidote to Iannaccone's list (Kwon, Ebaugh, and Hagan 1997). Members of the Houston Korean Church consist largely of recent immigrants to the United States, many of whom arrive penniless, homeless, friendless, and unable to speak English. By joining the church they are immediately welcomed into a social network that offers friendship, help, and solid opportunity. As an ethnically based church, it can cater to the special needs of its members. New arrivals are aided in buying a vehicle, finding housing, obtaining job referrals, baby-sitter referrals, Social Security information, and translating services, making airport pickups, making and receiving visitations for new babies and hospitalized members, registering children for school, applying for citizenship, and dealing with the courts; the list of material benefits goes on and on.[1] New members frequently are employed by more established members who own businesses and who also gain from the arrangement. The following autobiographical account summarizes the tremendous material benefits received by one member:

> When I came to Houston, I did not know a single person here. I had
> only about $200 in my pocket. As I arrived, I went to a Korean church. I
> knew that the church was able to find me a job. Soon, they found me a
> position in a restaurant which was operated by a church member. He al-
> lowed me to eat as much as I wanted and to sleep in his restaurant at
> night . . . that's how I saved the money to start my "road sales" business.
> I continued to attend the church. Later, when I opened my shop, many
> church members came to my shop as customers. (Kwon, Ebaugh, and
> Hagan 1997, 254)

It might seem that this person had only material benefits in mind, which provided all the incentives needed to join such a church. However, by now we should be suspicious of this conclusion. As we have seen, the practical side of religion does not negate the spiritual and metaphysical sides; they go together and reinforce each other. Members of the Korean Christian Church obtain psychic benefits that are regarded as just as important as material benefits, including emotional support, a sense of belonging, and respect that is lacking from their position in American society. Here is a description of Mr. Kim, who came to America before the Houston Korean Church existed but who nevertheless joined as an already successful businessman:

> Mr. Kim, who came to Houston from Korea 20 years ago, said that when
> he was working from 5 A.M. to 11 P.M. seven days a week, he simply did
> not have any time to ponder his emotional needs. After he had estab-
> lished himself as a prominent businessman in Houston, he started to feel
> depressed and bored. He began to attend the Korean Christian Church
> and become part of a cell group that, as he described, "saved" him from
> his emotional problems. Through the church, he found friendship, a sense
> of belonging, and "the reason for being," which he had forgotten since his
> arrival in America and during his struggles to establish himself. (252)

Mr. Kim may also have profited financially from church members who became employees and customers, but I am prepared to accept his testimonial at face value. I do not think that people are driven entirely by material interests, much less material self-interest (Sober and Wilson 1998; Wilson and Sober 2001). I find it plausible that Mr. Kim eagerly shared his wealth with others and traded dollars and cents for a sense of belonging. The need for respect is especially poignant. Mr. Son, a sixty-eight-year-old deacon of the church, was a college professor in Korea but was turned away by American universities and ended up opening a flower shop. I find it plausible that the respect accorded to him as church elder was more important to him than material benefits. He may well have become poorer to earn respect. In part because respect is a basic human need, especially valued by Koreans and in especially short supply for Korean immigrants, the Houston Korean Church includes a large number of "official" positions that grant a formal status to their holders.[2]

The term "cell group" in the above cited passage refers to a novel social organization called a cell group ministry that was developed only a few years ago by two ministers in South Korea. Church members live throughout the Houston metropolitan area. In addition to weekly Sunday services for the entire congregation at the main sanctuary, members also meet monthly in neighborhood groups of twenty to twenty-five people (the cells). Each cell has two leaders in charge of coordinating group activities, and the meetings are held on a rotating basis in the houses of its members for worship, food, and good company. New members are assigned to a cell, which becomes their family and is responsible for helping them meet the challenges of their new life. Thus, the church remains vibrant at the level of small-scale social interactions even after the total congregation has grown large as a church.

The cells are integrated into the larger church by a number of ingenious methods. The Houston area is divided into five regions with a minister of visitation and an elder assigned to each region (recall that Calvin's Geneva was similarly divided into sectors). Cell membership within each region is shuffled annually to expand each member's social network and to prevent cells from competing with each other. The most ingenious innovation is a mailbox system located at the entrance to the main sanctuary of the church. Every member has a mailbox, and the mailboxes for each cell are arranged in a separate cluster. Not only do the mailboxes serve the obvious function of getting information from the church to its members, but they provide a method of monitoring and control. If a bulletin remains in a mailbox, church administrators, cell group leaders, and cell group members know that the person did not attend Sunday services. The member is usually called the next day and the bulletin is mailed to his or her home along with a tape of the sermon. In one deft stroke, the church quickly helps members in distress (one reason for being absent) and also makes a public event out of waning commitment (another reason for being absent).

The cell group ministry works so well that it is being considered for adoption by other Christian denominations, just as the social organization of Calvin's church was copied five hundred years ago. However, it does have the effect of preventing Korean immigrants from being fully assimilated into American society. Perhaps for this reason, it is less popular among second-generation Koreans, including this young man, who describes the religious involvement of his parents:

> Church, home and work, that's all they have. That's their whole life. They go to work in the morning, come home at night, go to church or talk to someone they know from church in the evening, and so on and on. They don't have any motivation to know or even a notion in their head that there is a whole world out there, the world that they don't know about and their children need to know about. (255)

Thus, the very same church that provides opportunities for recent immigrants can limit opportunities for their offspring, a point to which I will shortly return.

This wonderful example is full of lessons for our more general evaluation of the modern social scientific literature on religion. First, it shows that we don't need to reach back five hundred or two thousand years to find churches that provide solid net benefits for their members in purely material terms. Second, it suggests that material benefits are not the whole story. People do have psychic needs, but those emphasized by Kwon et al. (emotional support, a sense of belonging, and respect) are different and more familiar than the "religious satisfactions" listed by Iannaccone. Third, it shows that benefits and the degree of strictness required to secure them depend on the circumstances. The Houston Korean Church is not very strict, despite its clever mailbox system. It doesn't need to be, because its members already lack other options. Yet, the benefits that are so valuable to recent immigrants become cheap for their offspring. Even the psychic benefits can be obtained elsewhere. Strictness is not required to retain the parents, and no amount of strictness could retain many of the offspring. This kind of context-dependence cannot be ignored in the study of religion.

The benefits provided by this particular church (for the parents) seem so obvious that they don't need to be measured. However, suppose a skeptic insisted that we actually measure the benefits. We therefore go to the Houston bus station and follow the next one hundred Korean arrivals. Fifty of them are penniless and have no social connections. The other fifty are also penniless but have social connections unrelated to the church. Perhaps their relatives in Korea have business interests in Houston and have already arranged for apartments and employment opportunities. All one hundred are good cost-benefit reasoners with the same material needs and familiar psychic values. The fifty without social connections quickly decide to join the church. Some of the remaining fifty also decide to join the church, but others elect to make use of their other social connections instead. As scientists, we dutifully compare the material and psychic welfare of those who joined the church with those who did not.

Of course, this is not the right comparison. If our one hundred Koreans are really shrewd, the nonmembers will become more prosperous than the members; after all, they had the option to join and refrained. The appropriate comparison is more subtle. We must compare the material and psychic welfare of individuals who join the church with a ghost category of what their welfare would have been had they not joined. This can be a challenging task, even when the difference is between prosperity and destitution. Perhaps we can find a comparable city that receives Korean immigrants but does not have a church so tailored to their needs. Perhaps we can measure the welfare of Korean immigrants who came to Houston before the church was formed. Perhaps we can rely on the fact that some Houston immigrants fail to hear about the church. Care is required to make the right comparisons when people are intelligently choosing their options on the basis of varying constraints, especially when many of the constraints are unknown to the scientific investigator.

My analysis of the Houston Korean Church sounds like that of a hard-headed economist who cares only about the bottom line of utility maximization. However, it is a world apart from the economic approach to religion in its current form. This problem rests ultimately with rational choice theory, which is powerful at predicting the consequences of generic utility maximization but weak at predicting the specific content of the utility. Plug in material benefits and a few familiar psychic values and rational choice theory goes where I have taken it. Plug in other utilities and it goes elsewhere. Returning to the nautical imagery that began chapter 2, rational choice theory is like a mighty fleet of battleships with a lousy sense of direction. To bring the fleet under control, we need to know something about the specific content of utilities. It is here that evolutionary theory can succeed where rational choice theory has failed.

WHAT PEOPLE WANT

It may seem audacious or even offensive to claim the ability to predict what people want—in economic jargon, the utility they attempt to maximize. People want to decide for themselves what they want and not to be told by someone else! Without violating this fierce instinct for free choice, which I would say dates back to our hunter-gatherer days, some very general principles can be established. Let's begin with the adaptationist's fantasy of a species perfected by natural selection. Every choice made by an individual enhances fitness compared to its other options. It doesn't matter whether this species evolved by group or individual selection; in the first case individuals choose unerringly to contribute to group fitness and in the second case to their own relative fitness within the group. The important point is that every choice can be explained in terms of the "hard" currency of survival and reproduction, appropriately defined. The "soft" inner world of psychic costs and benefits can be ignored.

The relationship between "soft" and "hard" in this example, the "inner" world governed by psychic forces and the "outer" world governed by survival and reproduction, is identical to the distinction between "proximate" and "ultimate" in evolutionary theory discussed in chapter 2 (see also part II of Sober and Wilson 1998). Individuals of our fantasy species have an inner psychological world that actually causes their behavior in a proximate sense, but it has been so thoroughly designed by natural selection that survival and reproduction in the outer world is sufficient to predict the behavioral output of the inner world. For example, suppose that individuals survive better in groups than alone, a fact that we can readily determine by observation or experiment. Natural selection has endowed individuals with a psychological mechanism that causes them to join groups, which they experience as a warm feeling of fellowship. We would not say "the individual joins groups because it survives better and because it receives a warm feeling of fellowship," as if these two benefits were similar in kind and can simply be added together. In proximate terms the individual joins groups only because of the warm feeling of fellowship, while in ultimate terms it joins groups only because survival is enhanced. The proximate/ultimate distinction is one of the most fundamental that can be made in evolutionary theory and must be born in mind when thinking about religion.

The fact that the behavior of our fantasy species can be predicted without reference to its inner world is a blessing because the inner world is so difficult to study. Nevertheless, that does not make it unimportant or uninteresting. A properly working inner world is assuredly important because individuals would not behave adaptively without it. Give the animal a drug to remove the warm feeling of fellowship and it will wander alone into the mouth of a predator. Biologists have entertained themselves for decades working out the wonderful anatomical, physiological, and molecular mechanisms that enable organisms to survive and reproduce. Proximate psychological mechanisms should be studied with the same interest and wonder. Why call them dull just because they work well? Furthermore, to the degree that real species depart from our fantasy species, the inner world of psychological mechanisms will not do its job to perfection and must be understood to predict the nonadaptive behaviors that result. When rational choice theorists talk about "soft" psychic benefits that do not produce "hard" material benefits, they are implicitly assuming that people are behaving nonadaptively in their environments.

The adaptationist program

People *do* behave nonadaptively in their environments, both as individuals and as groups. We are not the fantasy species outlined above. However, we do not know how far we depart from the fantasy species. Worse, we know so little about the proximate mechanisms that motivate our behavior that we are reduced to pathetic guesses such as "warm feeling of fellowship." Like it or not, the best information we have concerns the outer world of survival and reproduction. This information is sufficient to predict behavior at the adaptive extreme of the continuum, but it becomes progressively meaningless as we proceed to the nonadaptive extreme. How can we make intellectual progress in this situation?

This is a familiar problem for evolutionary biologists, as we saw in chapter 2. No species is perfectly adapted to its environment, and the causes of nonadaptive behavior are multiple and relatively difficult to study. A practical two-step solution is to first predict behavior purely on the basis of adaptation, knowing that the prediction is unlikely to be exactly correct. The hope is for it to be approximately correct, to the degree that the organism has evolved to behave adaptively in its environment. Departures from the prediction are also instructive by pointing to specific causes of nonadaptive behavior.

The second step is to identify the proximate mechanisms that evolved to produce adaptive behavior, whose failure under various circumstances can also explain the observed nonadaptive behavior.[3] Some of my evolutionary colleagues make a game of asking people what they want from life, and never yet have they received the answer "I want to maximize my fitness," or even a rough equivalent. Evolution has endowed us with a number of subsidiary drives that tend to enhance fitness but which we regard as psychological ends in themselves. The sex drive is usually mentioned to illustrate this point. Sexual pleasure is nature's way of causing us to have babies. We have a burning desire for sex, not babies. If you give men vasectomies so that sex does not lead to babies, they still want sex, which is a psychological end in itself. This example is complicated by the fact that people sometimes do crave babies and sex can have functions other than reproduction, but the point it illustrates stands firm.

What other drives can we imagine to be psychological ends that reliably serve as means to the evolutionary end of survival and reproduction? One good candidate is a drive to become a respected member of a strong supportive group. To be an outcast was a virtual death warrant for our hunter-gatherer ancestors. We therefore probably evolved a deep yearning for belonging and respect, regardless of their present-day consequences. Craving respect when it doesn't lead to material benefits is no more puzzling than craving sex when it doesn't lead to babies.

The two-step procedure that I have outlined is the adaptationist program in evolutionary biology in the pragmatic sense of the term; not a claim that everything is adaptive but a method for asking questions when an adaptationist hypothesis can be framed much more precisely than alternative hypotheses. Of course, the more we can formulate precise nonadaptation hypotheses, the more adaptation loses its special status in the explanatory framework.

Using the adaptationist program to study religion

The adaptationist program provides a framework for studying religion that is quite novel against the background of rational choice theory and the rest of the social sciences. It begins by dividing the study of religion into two separate enterprises, which can be labeled prediction and production.

Prediction is the study of adaptive behavior with minimal reference to proximate mechanisms. This enterprise is like rational choice theory, except that the utilities are less arbitrary. Costs and benefits are limited as much as possible to the hard currency of survival and reproduction. It is redundant to add psychic factors to the utilities because they are designed to produce the material consequences that have already been included. Proximate and ultimate explanations cannot be mixed, as we saw in the case of our fantasy species. We hope that our adaptationist predictions will succeed sufficiently often to pay their way, but we also expect them to fail occasionally, since we are not, after all, the fantasy species. When our predictions fail, we try to use the failures to identify the causes of nonadaptive behavior, such as proximate mechanisms that evolved because they increased fitness on average, even if they decreased fitness in our particular case. Thus, if we see individuals spending hard-earned resources for something like respect or a sense of belonging, which does not seem to offer a material return on their investment, we should first acknowledge that our adaptationist prediction has failed but then claim that the explanation is not far to seek. A conservative approach would be to see how often religious groups deliver net benefits to their members in purely material terms, measured ultimately in survival and reproduction, adding psychic costs and benefits to the utility only when suggested by the data. I did not begin by assuming that respect is valued for its own sake in my analysis of the Houston Korean Church. It was suggested by the data.

To summarize, the prediction arm of the adaptationist program attempts to explain religious groups as adaptive units purely in terms of survival and reproduction (ultimate causation), ignoring psychic factors and other proximate mechanisms as much as possible. I don't mean to underestimate the difficulty of this enterprise, but it is vastly more disciplined and justified on the basis of first principles than the "anything goes" content of utility in rational choice theory (Hirshleifer 1977).

Production refers to the mechanisms that actually cause behavior in a proximate sense. These mechanisms are not required to predict adaptive behavior but they are assuredly required to produce it. By analogy, suppose that you purchase some tax-preparation software for your computer. You know what it is designed to do in minute detail without knowing anything about the code in which it was written, but a single error in the code can prevent the program from producing its advertised features. A complete understanding of the software requires understanding its function and its code, especially if there is an error that needs to be fixed.

Before describing the production arm of the adaptationist program, let us see how rational choice theory approaches the same enterprise. The first three propositions in Stark's formal theory of religion are statements about how people think: People attempt to make rational choices, they attempt to formulate explanations, and they attempt to evaluate explanations on the basis of results. Iannaccone concurred in the passage quoted in chapter 2; he said that the basic issue is whether people attend to costs and benefits and act so as to maximize their net benefits. The principal alternative, according to Iannaccone, is "unreflective action based on habit, norms, emotion, neurosis, socialization, cultural constraints, or the like—action that is largely unresponsive to changes in perceived costs, benefits, or probabilities of success."

In this fashion, rational choice theory is presented as a theory of proximate mechanisms—a theory that tells us how people think. Furthermore, the theory can be presented in a remarkably few statements. Just as Newton could explain so much about the physical world with his three laws of motion, a few law-like statements about rational choice suffices to explain the length and breadth of human psychology, secular and religious. Unfortunately, rational choice theory doesn't take its own mission seriously. All it really wants is justification for making predictions on the basis of generic utility maximization. When pressed, rational choice theorists easily back away from their specific assumptions about human psychology. As long as people behave as if they are cost-benefit reasoners, then it doesn't matter how they actually think.

In short, rational choice theory often seems to be engaged in what I have called the production enterprise—figuring out the psychological mechanisms that cause behavior in a proximate sense—when in fact it is engaged almost entirely in the prediction enterprise—figuring out how people behave on the basis of generic utility maximization. All the talk about how people think is just a metaphor that justifies the utility maximization principle. When judged as a theory of proximate mechanisms, rational choice theory fails miserably. Its law-like propositions are surely important elements of human psychology, but they are neither necessary nor sufficient to predict behavior in particular cases. Furthermore, the long list of factors that are rejected in favor of rational choice—norms, emotions, functionalism, and the like—have always included cost-benefit reasoning within their own structure. Durkheim assumed that people evaluate religion on the basis of its secular utility, which is why it is so functional for society. Most discussions of norms assume that they are obeyed to avoid punishment (e.g., Cialdini and Trost 1998). Emotions are required for rational thought, since alternative courses of action have no costs and benefits to be compared unless they are endowed with emotional significance (Damasio 1994). The long list of factors that Iannaccone rejects in favor of rational choice theory strikes me as more an admission of defeat than a proclamation of victory, like someone throwing away a box of model airplane parts out of frustration because he can't make the pieces fit together. A proper theory of proximate mechanisms must take back the parts that Iannaccone rejected—habit, norms, emotion, socialization, cultural constraints, and more—and show how they interact with each other and with cost-benefit reasoning to (partially) produce adaptive behavior.

The adaptationist program does not require a psychological metaphor to justify the principle of utility maximization. It has the process of natural selection, whose product, adaptation, is utility maximizing by definition. Liberated from the psychological metaphor, the production arm of the adaptationist program is free to explore the real proximate mechanisms that cause religious groups to function as adaptive units. Perhaps *Homo economicus* will emerge as the final answer, but perhaps he will not.

How should we begin to understand the proximate mechanisms that enable religious groups to work as well as they do? By seeing if traditional statements about religious belief can be accepted at face value. Evans-Pritchard, who criticized Durkheim in other respects, credited him with an insight into "a psychological fundamental of religion: the elimination of the self, the denial of individuality, its having no meaning, or even existence, save as part of something greater, and other than the self." Of course, we don't need scholars like Durkheim and Evans-Pritchard to make this point, which can be learned directly from countless religious texts and people, from the Buddha to Calvin.

Multilevel selection theory allows us to explain this psychological fundamental of religion at face value, as one of the most important proximate mechanisms that evolved to enable groups to function adaptively. A group of people who abandon self-will and work tirelessly for a greater good will fare very well as a group, much better than if they all pursue their private utilities, as long as the greater good corresponds to the welfare of the group. And religions almost invariably do link the greater good to the welfare of the community of believers, whether an organized modern church or an ethnic group for whom religion is thoroughly intermixed with the rest of their culture. Since religion is such an ancient feature of our species, I have no problem whatsoever imagining the capacity for selflessness and longing to be part of something larger than ourselves as part of our genetic and cultural heritage. The only reason that selflessness ever appeared unreasonable was the free-rider problem. As we have seen many times over, free-riding is a problem for which there are solutions. The categorical rejection of selflessness on the basis of the free-rider problem will be seen in retrospect as one of the great oversimplifications of all time. However, the free-rider problem does mean that religion cannot be based exclusively on unconditional selflessness. Religion is selflessness with strings attached. My analysis of Calvinism was an attempt to describe the true psychology and culture of a religion, strings and all.

The fact that selflessness is so central to religion helps explain why religious people themselves often deny its utilitarian nature. Altruism is psychologically paradoxical. Those who care least about themselves prosper most, as long as they live among like-minded individuals. The prize of personal welfare is obtained by running directly away from it. In principle, it should be possible for a psychological egoist to grasp this fact and behave tirelessly on behalf of others to maximize her own gain. In practice, it often works better to drop the principle of psychological egoism altogether—abandon self-will—and make the welfare of others a more primary value. It is therefore unsurprising that some religious believers reject the notion of religion increasing their own prosperity as an average member, since that is the very opposite of their own motives.

Similarly, one frequently hears that religion goes beyond helping others. In his popular history of world religions that is an inspirational work in its own right, Smith ([1958] 1991) repeatedly says that religion accomplishes community but is about much more than community. This statement might be true at the proximate level but not at the ultimate level. Loving and serving a perfect God is vastly more motivating than loving and serving one's imperfect neighbor. Mother Theresa was able to work tirelessly for the poor of Calcutta in part because she saw God in their wretched bodies (Kwilecki and Wilson 1998). Nevertheless, these exalted feelings may have evolved, both biologically and culturally, precisely because of their utilitarian consequences. The same can be said for romantic love. Couples who fall in love don't think "this is my best prospect for survival and reproduction." Nevertheless, the rapture they feel probably evolved for exactly that purpose. I am aware that lovers and religious believers alike might object to putting the utilitarian cart before the psychological horse, and I will return to this subject in chapter 7.

To summarize, the production arm of the adaptationist program accepts traditional religious belief in some respects (abandonment of self-will) and challenges it in others (the ultimately utilitarian nature of religion). What it does not do is shoehorn the psychology and culture of religion into a narrow conception of self-interested utility-maximization. Instead, it admits the full list of factors that superficially are excluded from rational choice theory and uses functional thinking to suggest how they are assembled into an integrated system for motivating adaptive behavior. For example, Calvin attempted to instill an unquestioning faith in God's will, which is the suspension of cost-benefit reasoning. However, he provides a logical argument for faith that employs cost-benefit reasoning. The argument relies upon emotions such as fear and joy. Children are socialized to accept the belief system at an early age, and free-riders are excluded by a well-enforced system of norms. Cost-benefit reasoning is merely one component of this complex adaptive system, whose properties can be deduced from the purpose for which it is designed, but could never be deduced from cost-benefit reasoning alone.

THE ADAPTIVE UNIT AS A PRIVILEGED UNIT OF STUDY

So far I have shown that evolutionary theory can go beyond rational choice theory (and other social scientific approaches to religion) in predicting the specific content of utilities and in distinguishing between proximate and ultimate causation. In addition, evolutionary theory can provide some commonsense methods for studying religion that may appear obvious in retrospect but are seldom employed in religious research. In chapter 3 I mentioned that adaptations must be studied at the appropriate scale to be seen at all. The fact that upstream and downstream guppies are locally adapted to their environments would have been missed if we regarded all guppies as a homogenous unit. Imagine an ecologist who failed to distinguish among bees, wolves, and trees in her study of nature. Clearly the many features that adapt these very different organisms to their environments would appear meaningless. This example may sound absurd, but studies of religion commit the same kind of error all the time by aggregating many different religious denominations, each of which consists of many congregations—into homogenous categories.

One reason that I have been able to provide such vivid portraits of religious groups as adaptive units is that I identified roughly the appropriate scale. I was like an ecologist who studies bees rather than bees-wolvestrees. The immense functionality of the water temple system of Bali would have been obscured even by lumping it with other temple systems in Bali. The functionality of the Korean Christian Church would have been obscured by lumping it even with closely allied denominations whose members face different needs.

To see how social science research on religion often fails to make these distinctions, consider two recent papers on religious involvement and social ties (Ellison and George 1994; Bradley 1995). Recall that the Korean Christian Church helped its members primarily by providing social ties.

When this single congregation was studied by itself (the appropriate scale of analysis), it obviously offered its members more and better social ties than their meager options available elsewhere. It passed the net material benefit test with flying colors. Working with survey data, Ellison and George examined the relationship between religious involvement and social ties in a sample of 2,956 households in the southeastern United States. Number of nonkin ties was measured by items such as "Other than members of your family, how many persons in this area within one hour's travel do you feel you can depend on or feel very close to?"—with response categories ranging from "zero" to "ten or more." Religious involvement was measured as frequency of attendance at "church or other religious meetings," with response categories ranging from "never" (14.8 percent of the sample) to "more than once a week" (13.0 percent). As usual for research of this kind, a number of control variables were included such as age, gender, urban vs. rural, education, marital status, employment status, and household income. The results showed a positive relationship between religious involvement and both the number and quality of social ties. The average person who attends church several times a week enjoys roughly 2.25 more nonkin ties than the person who never attends. A positive relationship also exists with weekly telephone contacts, receiving gifts, business or financial advice, and help in performing home maintenance and repair tasks. Other potential benefits of social ties did not correlate with religious involvement. Bradley (1995) tested the same general hypothesis with a larger survey that encompassed the entire United States and came to similar conclusions.

These results could have been different. It is entirely possible that religious involvement cuts social ties outside the church even more than expanding them within the church, decreases the average quality of social ties, and so on. Thus, these studies broadly support the hypothesis that religion delivers net material benefits to its members in the form of social ties. Nevertheless, the results lack force and would do little to convince a skeptic. Ellison and George themselves begin the results section of their paper with the following caveat: "As is frequently the case in survey research, the explanatory power of these various models is modest, generally accounting for 10–18% of the variation in social resources" (1994, 54). So, what does it mean to have 2.25 more nonkin ties or 0.85 additional phone contacts per week—an extra dish at the church social? Are the benefits of religious involvement really so wimpy? Or has the power of at least some churches to benefit their members become invisible due to the scale of analysis employed? As we saw in the case of the one hundred Koreans, the signs of the correlations could even have been reversed, with religious involvement seeming to decrease the average quantity and quality of social ties, and I could still plausibly argue for the power of churches for those who chose to become members. The scale of analysis is simply too large for the questions we need to ask. Potentially strong evidence for religious groups as adaptive units is obscured by a fog of aggregation.

A similar story can be told for religion and deviant behavior. One would certainly think that the Ten Commandments, constant injunctions to follow the Golden Rule, and plain old-fashioned social control would cause church members to be more law-abiding than those who stay at home on Sunday. It therefore came as a surprise when Hirschi and Stark (1969) found no correlation whatsoever between measures of religious participation (Sunday school and church attendance as well as belief in hell) and delinquency among high school students in Richmond, California. Critics of religion, especially numerous among social scientists, chortled with glee at this exposure of hypocrisy that they had suspected all along. Yet, however severely I may criticize Stark's formal theory of religion, he is a consummate scientist who respects the subject he studies. He and his colleagues forged ahead and ultimately documented a more sensible, if complex, relationship between religion and deviant behavior. Their main unit of analysis was the Standard Metropolitan Statistical Area (SMSA), for which rates of deviant behavior are tabulated by the Bureau of Vital Statistics. Church membership data is also available for these units, which varies from 96.6 percent of the population in Provo, Utah, to 25.0 percent in Eugene, Oregon. Using this nation-wide data base, Stark and Bainbridge (1997) demonstrate a strong negative correlation between the proportion of people in a metropolitan area that are enrolled as members of a church and rates of deviant behavior ranging from suicide, to crimes against property and other people, to so-called victimless crimes such as drug and alcohol abuse. Religion appears to be important apart from social integration, which can also exist in a nonreligious context. However, the particular denomination, such as the proportions of Catholics vs. Protestants in a metropolitan area, appears to have no effect (with some exceptions noted below). Another major result is that actually participating in a church appears to be more important than personal feelings of religiosity.

Once again, these results broadly support the theory that religious groups function as adaptive units in purely secular terms. In fact, despite a withering attack on Durkheim earlier in their book, Stark and Bainbridge agree that the most religious metropolitan areas "exemplify the moral community as Durkheim conceived of it—a community integrated by shared religious beliefs which sustain conformity." Here, as in his analysis of early Christianity, Stark turns his rational choice battleship in a functionalist direction and hits his target with impressive force. If only he would keep the battleship pointed in the same direction!

Despite the broad support for the hypothesis that religious groups function as adaptive units, major questions cry out for answers that are obscured rather than enlightened by the scale of analysis. As we have seen, group selection often does not eliminate conflict but merely elevates it to the level of between-group interactions. The strongest prediction is that churches cause their own members to behave prosocially toward each other. Predicting conduct toward outsiders is more complex, yet the crime statistics do not distinguish between within- vs. between-group interactions. Another major question concerns the importance of personal religious belief. Religions go to such lengths to arrange the furniture of the mind that it is difficult to believe that personal religious beliefs—a misnomer, since they are so largely a property of the church—have no effect on prosocial behavior. They may be insufficient without an accompanying social organization, but the reverse might also be true. Alcoholics Anonymous and similar organizations provide a good example. Members who join these organizations assuredly need help as far as their personal welfare is concerned, yet they are told that self-interest is the root of their problem:

> Selfishness—self-centeredness! That, we think, is the root of our troubles.
> Driven by a hundred forms of fear, self-delusion, self-seeking and self-
> pity, we step on the toes of our fellows and they retaliate. Sometimes
> they hurt us, seemingly without provocation, but we invariably find that
> at some time in the past we have made decisions based on self which
> later place us in a position to be hurt. . . . Above everything, we alcohol-
> ics must get rid of this selfishness. We must, or it kills us! (*Alcoholics Anon-
> ymous* 1976, 62)

AA works hard to instill belief in a higher power, abandonment of self-will, and service to others as essential steps toward personal recovery. Scientific research is required to determine the true efficacy of these personal beliefs. Can the survey data on alcohol abuse and religion analyzed by Stark and Bainbridge (1997) provide the answer? It turns out that there is a strong denominational effect on alcohol abuse, unlike other forms of deviance. Conservative Protestant denominations discourage drinking more than liberal Protestant denominations and the Catholic Church. Although the Catholic Church obviously does not condone alcoholism or any other self-destructive behavior, in practice it frequently does not limit alcoholism as effectively as conservative Protestant denominations. In fact, according to Stark and Bainbridge, regions of the United States with a high proportion of Catholics never even bothered to enforce the prohibition laws during the 1920s. Sobriety is one purely secular benefit that the Catholic Church often does not provide for its members, with consequences that can be measured in the hard currency of mortality, such as deaths from cirrhosis (83).

Stark and Bainbridge tell a fascinating story about religion and substance abuse. They are virtuosos when it comes to survey research and are sensitive to what they call the "ecological fallacy"—the attribution to individuals of findings based on aggregate units. They take care to show that Catholics and liberal Protestants really do drink more than conservative Protestants as individuals. All of these results are important in their own right, but they don't touch the question of whether alcoholics aid their recovery by trading the psychology of self-interest for the psychology of religious belief in the context of a program such as Alcoholic's Anonymous. Scientific research at a finer scale is required to answer this particular question (Hart 1999).

I encountered this problem again and again in my perusal of the modern social scientific literature on religion. Multiple streams of evidence seem to support the theory that religious groups function as adaptive units, yet it is often hard to tell because the streams are shrouded in a dense fog of aggregation. Science is supposed to sharpen our perception, not dull it, and I found myself yearning for religion to be studied in the way that evolutionary biologists study the rest of life: with a primary focus on the units that are adapted to survive and reproduce in their environments. That is why case studies of single religious groups that I did encounter, such as the Korean Christian Church, were so refreshing despite their qualitative nature. Of course, studies of single religious groups need not remain qualitative. It would be wonderful to put a number on the net material benefits received by members of the Korean Christian Church, compared to their other options. It would be equally wonderful to calculate net material benefits for members of a Catholic parish, in which the church's relative ineffectiveness at preventing alcoholism is carefully entered on the cost side of the ledger. Our purpose is not to show that churches invariably benefit their members but to measure their costs and benefits and to explain the resulting patterns in terms of our theoretical framework. The essential information must be gathered on a congregation-by-congregation basis before congregations can be aggregated for a more general analysis.

In my review of the modern social scientific literature on religion, I did encounter one stream of evidence that shines through the fog of aggregation with special clarity. It involves the relationship between cults, sects, and churches, which has been proclaimed as one of rational choice theory's greatest explanatory triumphs.

THE LIFE CYCLE OF DENOMINATIONS

Religious denominations range from huge established churches that encompass most of the population, to tiny sects that reject the larger churches as corrupt and regard themselves as keepers of the original faith, to cults that invent or borrow from elsewhere a set of religious beliefs that is completely different from those of either churches or sects. In America, the United Church of Christ is a church, the Jehovah's witnesses are a sect, and Zen Buddhism is a cult even though it is a dominant religious tradition elsewhere.

Long ago, religious scholars noticed that these three categories are related to each other in a systematic way: The churches of today are the sects and cults of yesterday. For example, the United Church of Christ was once called the Congregational Church and was the religion of the Pilgrims, who were so at odds with the Church of England that they migrated to the New World. Religious denominations seem to have a life cycle. They begin as sects or cults, grow into churches, give rise to offspring sects, and then mysteriously senesce, to be replaced by their own offspring or by new cults. Actually, we should be cautious about employing the organismic metaphor in this case. I expect churches to be like organisms in the sense of being well adapted to their environments. Churches try to reproduce by creating new congregations of their own kind, which is described as church growth in the religious literature. They are not necessarily designed to create new sects that differ from themselves. They certainly do not nurture new sects in any way resembling parental care, although they do nurture their own new congregations. Finally, senescence is still poorly understood from an evolutionary perspective in real organisms, much less churches. Rather than carelessly employing the organismic metaphor, we need an explicit hypothesis to describe the cycle of religious growth and decline that has taken place around the world and throughout history.

Rational choice theory claims to offer such an hypothesis. Alas, as we have already seen, rational choice theory is so lax about its definition of utility that it fails to distinguish among three major hypotheses:

Hypothesis 1: Group Benefits. According to this hypothesis, the primary purpose of religion is to provide collective benefits for its members, measured ultimately in survival and reproduction. Religion is needed most by people who do not already have abundant resources and who have the most to gain from collective action—the poor rather than the rich. However, to the extent that religion achieves its primary goal, it makes poor people rich and therefore fundamentally alters their religious needs. It is an overstatement to say that rich people have nothing to gain from collective action. However, a religion designed to provide collective benefits for the wealthy will be very different than a religion designed to provide collective benefits for the poor. Thus, a sect or cult doesn't senesce as it grows into a church; it adapts to the changing wealth of its members caused by its own success. John Wesley (1976, 9:529), founder of the Methodist Church, stated this hypothesis with crystal clarity:

> I do not see how it is possible, in the nature of things, for any revival of true religion to continue for long. For religion must necessarily produce both industry and frugality. And these cannot but produce riches. But as riches increase, so will pride, anger, and love of the world in all its branches.

It is easy to see the logic of this hypothesis operating for the Houston Korean Church described earlier. Imagine what would happen if the flow of immigrants dried up, leaving only the second generation of Korean-Americans who do not need any of the services provided to their parents. These members might still have something to gain from collective action, but nothing as crucial as the life-saving services offered their parents. Nor is it obvious that a church composed entirely of Korean-Americans can provide collective benefits for members who are eager to pursue the American dream. The church would have to scramble to prevent a massive decline in membership. It almost certainly would drop its cloying practice of calling Sunday no-shows on Monday. Instead of teaching its members English and helping them cope with American society, it might start teaching them Korean and helping them rediscover their ethnic roots. Yet these benefits are pale compared to those that could be offered only a generation ago. Since religion is fundamentally a solution to basic needs, no matter how well the church adapts to its own success, it will lose its original vitality and experience declining membership and commitment. A "strong" sect will become a "weak" church.

Hypothesis 2: Individual Benefit. The second hypothesis focuses on the unequal sharing of wealth within the church. As before, the primary purpose of religion is to generate collective benefits for its members. Religions almost invariably include safeguards against free-riding, including the control of leaders, as we saw in the case of Calvin's Geneva. Nevertheless, despite these safeguards, self-will often does prevail within churches, resulting in some members profiting at the expense of others or at least without sharing their wealth as demanded by the selfless spirit of religion. When a church becomes sufficiently corrupt, its more pious members, who by now have become its poorest members, leave to form a church of their own with renewed safeguards against free-riding. A lax church gives rise to a strict sect.

Isaac Bashevis Singer's account of Jewish communities in his novel *The Slave,* which I discussed in chapter 4, illustrates the logic of this hypothesis. Jacob is a genuinely pious man who honors the spirit of his religion. Many—perhaps most—of his brethren also practice selflessness to the best of their ability, but a few make a mockery of their religion by exploiting it for personal gain. These people find ways to bypass even the most powerful safeguards against free-riding, insinuating themselves into leadership positions and corrupting the spirit of religious law even while observing its letter in every detail. Purifying the church requires either purging it of its corrupt members or abandoning it to form a new church.

It is important to clarify the use of words such as "spirit," "corrupt," and "purify" in the preceding two paragraphs. All religious groups probably include some people who work diligently to achieve collective benefits and others who free-ride. If all religious groups include free-riders, shouldn't free-riding be included in the definition of religion? Not necessarily. Religious people clearly view religion as an ideal that is only approached by actual religions. The ideal religion completely eliminates self-will and the free-riding that comes with it. The fact that real churches only approach the ideal need not alter the definition of the ideal. By analogy, consider multilevel selection theory, which partitions natural selection into within- and between-group components. The ideal group selection model completely eliminates selection within groups. Most groups fall short of the ideal, but that does not require us to change the definition of group selection. In fact, the correspondence between "ideal" religion and "ideal" group selection is gratifyingly close. The meaning of religious-sounding terms such as "spirit," "corrupt," and "purify" can be accepted at close to face value in multilevel selection theory.

Size is a critical factor in the so-called corruption of churches. Small groups easily police their members, including their leaders, who are chosen from their own ranks and can be deposed just as easily. Small groups are also maximally sensitive to their own basic needs. As religious movements grow, they must achieve unity at a larger scale than before. Not only does social control become more difficult to impose, but collective action designed to benefit the church writ large does not always benefit each local congregation to the same degree. If congregations are interested primarily in their own basic needs, they may begin to feel used by their own church and tempted to break away to form their own sect. This dynamic can be seen very clearly in the history of the Quakers (Ingle 1994) in addition to many other religious movements.

Hypothesis 3: Religion as a Byproduct. The third hypothesis focuses on religion as a byproduct rather than as a group-level adaptation. The primary purpose of religion is not to enhance the material welfare of its members through collective action but rather to obtain what cannot be had through the invention of supernatural agents. The psychology at the heart of religion is hugely beneficial, but only in a secular context, where it succeeds in obtaining the basic resources that can be had. In a religious context, the economic mind is merely spinning its wheels uselessly to obtain what cannot be had. Far from becoming prosperous, religious groups should become poorer in their effort to trade real goods for promised rewards that ultimately are not forthcoming.

What, then, accounts for the systematic growth of sects and cults into churches, which in turn give rise to new sects and cults? Imagine a tiny sect of impoverished people, praying fervently for what they cannot have, from everlasting life to material benefits that are beyond their reach. The fact that these blessings cannot be obtained does not prevent them from being desired; hence the gods. However, there is nothing in this scenario to change the material welfare of the members, other than to make them a little poorer. If we check back during the next generation, they should be praying even harder, and for the goods sacrificed to the gods along with everything else. The first and second hypotheses relied on material prosperity created by collective action to explain the transition from cults and sects to churches. Byproduct theories cannot make use of this explanation, so what do they offer in its place?

Consider the religious needs of rich and poor individuals in a given population. The rich have everything money can buy and lack only what truly cannot be had, such as everlasting life. The poor lack what the rich have in addition to what truly cannot be had. It follows that the rich and poor will not feel very comfortable praying with each other. As Stark and Bainbridge (1985, 103) put it:

> It is very difficult for a religious organization to offer credible and effective compensators to some members for scarce rewards if it is at the same time exhibiting, in the lives of its most influential members, the legitimacy and availability of the rewards in question. For example, the most credible form of religious compensators for unfulfilled worldly desires is to reject worldly values. For people who are poor, the most effective compensators define material goods as of little value in comparison to heavenly bliss: self-denial on earth buys riches in the world to come.

My rendering of this passage is as follows. The best way for poor people to make themselves feel good is by denying the importance of worldly values, compared to the heavenly bliss that follows death. This belief does nothing whatsoever to increase the worldly prosperity of the believers. It merely makes them feel better. Unfortunately, this feel-good solution is spoiled by the presence of rich people in the same religious community. How these people became rich is not specified. For their part, rich people don't enjoy being envied and morally shunned by the poor, so the two types of people part company; the poor into other-worldly sects, the rich into this-worldly churches.

This is my best effort to formulate a church-sect theory based on religion as a byproduct rather than as a group-level adaptation. To convince the reader that I am not distorting the views of Stark and Bainbridge, consider the following passage:

> Our formulation of church-sect theory leads to the conclusion that no single religious organization can serve adequately the whole spectrum of needs and desires found in the religious marketplace. No society can survive with a wholly otherworldly church, for we must exist in this world and hence some people must be free of religious fetters to pursue worldly aims. Moreover, those most inclined to worldliness will have the power to create (or transform) religious organizations to suit their needs. But many other people deeply desire otherworldly comforts for their earthly dissatisfactions. This creates a social chasm no organization can straddle successfully. (1997, 111)

There can be no doubt from this passage that religion is being conceptualized as a "fetter" rather than as a wellspring of worldly benefits produced by collective action. The difference between a byproduct hypothesis and the adaptation hypotheses could not be more clear.

I regard a combination of the first two hypotheses as far more plausible conceptually and better supported empirically than the third hypothesis. However, I will not attempt to evaluate the empirical evidence in detail because I'm not certain that rational choice theorists, who know the evidence better than I, would disagree. The real problem with rational choice theory is not that it has committed itself irrevocably to a byproduct theory of religion, but that it wanders so aimlessly from one major conception of religion to another, from religion as a feel-good fetter to religion as required for survival and reproduction, as if these differences are not worth commenting upon. Even before we evaluate the three hypotheses, multilevel selection theory has done the considerable service of distinguishing among them and stressing the importance of knowing the difference.

If a combination of the first two hypotheses prevails over the third hypothesis, then all religious scholars, including rational choice theorists, should agree that the life cycle of religious denominations provides powerful evidence for the central thesis of this book: Around the world and across history, religions have functioned as mighty engines of collective action for the production of benefits that all people want.

LIFTING THE FOG FROM MULTIPLE STREAMS OF EVIDENCE

My original intention for this chapter was to see if the modern social scientific literature on religion supports the conclusions that we have already reached on the basis of traditional religious scholarship. The broad answer is "yes"; far from sapping the resources of the believer in a futile effort to get what cannot be had, religion more often enables the believer to get what can be had through the coordinated action of groups. This conclusion is based on multiple streams of evidence, as befits the quotation that begins the chapter. The life cycle of religious denominations, long a centerpiece of the rational choice school, provides especially strong evidence for group-level adaptation when properly understood.

The more familiar I became with the modern literature, the more this chapter acquired a second purpose; to show how evolution provides a framework for studying religion that differs from existing perspectives in the social sciences. I have no wish to steal rational choice theory's thunder or deny its many successes in the field of religious research. I hope my admiration for some aspects of the theory and even greater admiration for its practitioners are apparent. Nevertheless, the distinction between proximate and ultimate causation and identifying the appropriate scale of analysis are so basic to evolutionary theory that they cannot be ignored by the human social sciences. When these basic insights are reflected in research on religion, the evidence for group-level adaptation will probably emerge even more clearly.

Let me be the first to admit that the adaptationist program has not yet proven itself for the subject of religion. On the other hand, this book represents a three-year effort by one person. By comparison, the literature on guppies, which demonstrates the full power of adaptationism, represents hundreds of person-years of effort. When religious groups are studied this well from an adaptationist perspective, by social scientists and religious scholars who learn about evolution in addition to evolutionists such as myself who learn about religion, the hypothesis that religious groups function as adaptive units will either self-destruct—a virtue in science— or stand on very firm ground.

CHAPTER 6

FORGIVENESS AS A COMPLEX ADAPTATION

You have heard that it was said, "An eye for an eye, and a tooth
for a tooth." But I say to you, Do not resist an evil-doer. But if
anyone strikes you on the right cheek, turn the other also.

—Matt. 5:38–39

Thinking of religious groups as organisms encourages us to look for
adaptive complexity. The behavior of a well-adapted organism seldom
takes the form "Do *x*." More often it consists of a large number of if-then
rules; "Do *x* in situation *a*, do *y* in situation *b*, do *j* in response to individual
c but not individual *d*," and so on. In this chapter I will examine a single
religious concept—forgiveness—as a complex adaptation. To be sure, forgiveness is not an exclusively religious concept, but it is said to be the
hallmark of the Christian religion. Christian forgiveness is often described
by the single phrase "turn the other cheek," but this may be too simple,
like "Do *x*." What is the true nature of Christian forgiveness, and can it
be understood from an evolutionary perspective?

THE INNATE PSYCHOLOGY OF FORGIVENESS

Scientists often try to explain the world with the simplest possible models,
in spite of its inherent complexity. The simplest model for the evolution
of social behavior consists of a single population of individuals genetically
programmed to behave in one of two ways, altruistic or selfish, and who
interact randomly with each other in a pair-wise fashion. Pairs of altruists
outperform pairs of selfish individuals, while selfish individuals outper-

form their altruistic partner in mixed pairs. The individuals then reproduce in proportion to the benefits they have received from their interactions, creating an offspring generation in which the cycle is repeated. The grim Darwinian struggle continues until one type goes extinct or they settle into a stable equilibrium. It should be obvious from chapter 1 that this is a miniature group selection model in which the groups are the pairs of interacting individuals (see also Sober and Wilson 1998, chaps. 1 and 2).

The result of this minimalist model is not very encouraging for the evolution of altruism. The altruists go extinct unless the selfish individuals are really nasty to each other, in which case a stable equilibrium results in which each individual gets exactly nothing, on average, from its social interactions.

We must complicate the model to make it more interesting and to explore a greater richness of behavior. Two complications are to allow members of a pair to interact more than once and to allow them to change their behavior based on the history of their relationship. There are still only two behaviors—altruism and selfishness—but they can now be governed by an if-then rule, such as "start by behaving nicely but become selfish if your partner behaves selfishly." The if-then rules are inherited genetically and are pitted against each other in a grim Darwinian struggle, just like the behaviors in the original model. The final outcome is a single rule that takes over the population or a mix of rules in a stable equilibrium.

These more complicated models are still laughably simple, but they open a Pandora's box of possibilities. Even two behaviors can be built into an almost infinite number of if-then rules, so how do we know which to pit against each other? When in doubt, ask your friends. At least that is what political scientist Robert Axelrod did, by sponsoring a computer-simulation tournament in which anyone could submit a rule. The story has been told many times (e.g., Axelrod 1984) and will only briefly be recounted here. Even though many complex rules were submitted, which presumably would require big brains to implement, a very simple rule submitted by Anatol Rapoport called Tit-for-Tat (TFT) drove all the other rules extinct and won the tournament. TFT instructs the individual to start by behaving altruistically and thereafter to imitate the previous behavior of its partner. TFT remains altruistic with altruistic partners but quickly turns selfish toward selfish partners, although it does get burned during the first interaction. Since pairs of TFT behave nicely toward each other and take over the population, altruism becomes the only behavior expressed even though the capacity for retaliatory selfishness is latent in every individual.

Some readers may wonder why I am calling TFT altruistic when it is commonly described as a benign form of selfishness in the literature. After all, TFT wins the tournament, so how can it be called altruistic? This reasoning, which confuses selfishness with success, commits the fallacy of averaging across groups described in chapter 1. As multilevel selectionists, we must define altruism and selfishness on the basis of fitness differences within and between groups. Pairs of individuals are small groups, but they are groups nonetheless. TFT is constitutionally unable to beat its partner within its own group. It can only lose or tie. The reason that TFT wins the tournament is because pairs of TFT are more fit than pairs in which selfishness is expressed. TFT evolves by between-pair selection, not within-pair selection. Rapoport (1991) clearly appreciated this multilevel dynamic and himself regarded TFT as an altruistic strategy, even though many of his colleagues conceptualized TFT's overall success as a form of selfishness. As he put it, "The effects of ideological commitments on interpretations of evolutionary theories were never more conspicuous" (1991, 92).

TFT is often described as having three properties: It is nice (by beginning as an altruist), quick to retaliate (by immediately becoming selfish in response to selfishness), and equally quick to forgive (by immediately becoming altruistic in response to altruism). Thus, forgiveness—which many might regard as a uniquely human capacity—emerges as biologically adaptive in very simple evolutionary models of social behavior.[1] However, forgiveness succeeds only because it is tightly linked with retaliation. The two concepts are joined at the hip. A rule that instructs the individual to turn the other cheek all the time does not survive the Darwinian struggle, at least in the context of our tinker-toy model as it currently stands.

To make the model even more interesting, we must increase its complexity. Let's say that you are a TFT that occasionally makes mistakes. You intend to behave altruistically but now and then your behavior is interpreted as selfish by your partner. If your partner is also a TFT, during the next round of social interactions you will behave nicely toward him (based on his previous behavior) but he will behave selfishly toward you (based on his interpretation of your previous behavior). Then you will retaliate against his selfishness while he forgives you for changing your ways. The relationship has gone out of kilter, sending you and your partner into an endless cycle of alternating retaliation and forgiveness.

This seems like a silly conclusion, something worthy only of the Three Stooges. Real people obviously can resume amicable relations after one of them commits a faux pas. In the context of the model, however, we must confirm our intuition by rigorously specifying a new rule and showing that it beats all others in the Darwinian struggle for existence. That rule is called Contrite Tit-for-Tat (CTFT; Boyd 1989). It is exactly like TFT until it commits a mistake, when it accepts two selfish acts from its partner before retaliating. If you are a CTFT with a TFT partner and you commit a faux pas, you accept what he puts upon you during the next round of interactions without retaliating, causing him to revert back to niceness on the following round and bringing the relationship back into alignment.

The CTFT rule works only if individuals know they have made a mistake. What if they are unaware of their own faux pas? You are being perfectly nice as far as you are concerned when—Wham! For no discernible reason, your partner turns on you. The rule that ends up evolving in this situation is called Generous Tit-for-Tat (GTFT), which forgives selfish acts without retaliating a certain proportion of the time that depends on the frequency of mistakes. If it is too generous it risks being exploited by selfish partners. It if isn't generous enough it suffers from sour relationships with potentially altruistic partners. The equilibrium degree of generosity represents a balance between these two opposing forces (Boerlijst, Nowak, and Sigmund 1997).

Adding more complications makes our models of social behavior even more interesting. Let's say that you are like the original unconditional altruist who went extinct long ago but you have a way of proving your disposition to others. Perhaps you blush every time you tell a lie, making it impossible for you to lie convincingly. Your commitment to honesty makes you a valuable social partner, which can offset your vulnerability to selfishness (Frank 1988). Saintly unconditional altruism can evolve in this fashion, although it seldom takes over the whole population. In addition, another kind of commitment can evolve that seems like the polar opposite of altruism. Suppose that you steal a $100 briefcase from me, knowing that it will cost me more than $100 to get it back. If I care only about maximizing my own utility, you can steal from me with impunity. But suppose I am the sort of person who becomes so angry at being robbed that I will hunt you down to the end of the earth, no matter what my cost. If you are convinced of my vengeful nature, you will not steal from me in the first place!

Another complication involves increasing the size of the groups. Not all social behaviors take the form of pair-wise interactions. If an altruistic behavior such as a warning cry encompasses a wider circle of individuals than the actor and a single social partner, then this must be reflected in the model. Chapter 1 reviewed models of social behavior in larger groups, which need not be repeated here.

We have reached the end of our whirlwind tour of evolutionary models of social behavior. More leisurely tours are available elsewhere (Axelrod 1980; Dugatkin 1997, 1999). The most important conclusions for our purposes are these: First, these models explain social behavior purely in terms of reproductive success. Second, they begin at an extremely simple level. Third, the most fit rules for social behavior are primarily altruistic. They succeed by doing well as groups (pairs) rather than by exploiting their partners within groups. Fourth, the most successful forms of altruism are usually (although not always) protected from selfishness through the capacity to retaliate. Fifth, as we carefully develop these models, adding complications one by one, a profile of traits emerges that many would regard as uniquely human—niceness, retaliation, forgiveness, contrition, generosity, commitment, saintly unconditional altruism, and self-destructive revenge. It is amazing that such a human profile of traits emerges so naturally from such simple models based entirely on survival of the fittest.

These conclusions force us to revise long-standing attitudes about nature and human nature. The traits that seem so human are biologically adaptive and do not require a large brain. Why, then, should they be limited to humans? The capacity to forgive should extend far beyond us and our primate ancestors to any creature with sufficient brainpower to employ TFT-like rules. Even guppies have been shown to qualify (Dugatkin and Alfieri 1991a, b). They recognize their social partners as individuals, behave according to their own dispositions at the start of relationships, and become less altruistic in response to selfish partners. They try to associate with altruistic partners regardless of their own degree of altruism. Males even show off with daring acts of altruism in front of females (Godin and Dugatkin 1996). No one imagined such human-like behaviors in guppies thirty years ago, yet they are plausible in retrospect because they are adaptive and don't require that much brainpower.

Conversely, if these "human" traits are adaptive and widely distributed in nature, we need not attribute them to special conditions in humans. Our big brains might allow us to play the social game better than other creatures, but we are playing the same game. The days of thinking of our species as categorically different from the rest of the animal kingdom, at least in these respects, are over.

Evolutionary theories of human behavior are controversial, and I myself have criticized some current versions (Wilson 1994, 1999b, forthcoming). Nevertheless, I find it highly plausible that we are endowed with an innate psychology that is crudely approximated by the adaptive rules of the evolutionary models outlined above. We have an innate capacity for altruism, selfishness, retaliation, forgiveness, contrition, generosity, commitment, saintliness, and vengefulness.[2] More, we are endowed with a set of if-then rules for when to employ these traits. As I discussed in chapter 1, the word innate does not imply that our behaviors are entirely genetically determined. On the contrary, if-then rules cannot function without environmental information. Innate psychology facilitates and structures cultural evolution (the machine part of the Darwin machine), providing the building blocks with which innumerable cultures can be built.

To summarize, the basic capacity to forgive is a biological adaptation found throughout the animal kingdom, at least in a rudimentary form. It cannot be understood by itself but instead must be seen as part of a constellation of traits that function as a set of if-then rules. One of the strongest links is between forgiveness and retaliation. Forgiveness is often valued and retaliation devalued in everyday thought, especially Christian thought. For the moment, we need to suspend these value judgments. In the context of evolutionary models, retaliation is absolutely essential to keep the wolves of selfishness at bay. To retaliate can be divine.

PROXIMATE PSYCHOLOGICAL MECHANISMS

In chapter 5 I made a distinction between the prediction and production arms of the adaptationist program. We can predict the behavior of organisms largely on the basis of survival and reproduction, but these behaviors are produced by proximate mechanisms that also need to be understood. Science, the arts, and everyday experience confirm that the behaviors listed above are motivated by some of the most powerful emotions known to humanity. The thirst for revenge can drive people to desperate acts, and its satisfaction is described as sweet. Forgiveness is a cathartic event and can provide such relief that it has important health consequences (McCullough, Pargament, and Thoresen 2000). It has become standard fare for neurobiologists to think of emotions as an ancient set of mechanisms for motivating adaptive behavior that evolved far before our appearance as a species. Even our newfangled capacity for rational thought requires emotions, which provide the values upon which cost-benefit reasoning is based. The aptly titled book *Descartes' Error* (Damasio 1994) shows that thinking cannot be divorced from feeling. Brain-damaged patients who have lost their ability to feel literally cannot make decisions. They can contemplate alternatives but have no way to rank them.

Thus, the production arm of the adaptationist program converges nicely with the prediction arm. Powerful proximate mechanisms can be demonstrated for the behaviors that are predicted to be adaptive. However, the link between behavior and emotion needs to be interpreted with appropriate subtlety. Forgiveness can occur at the behavioral level but not the emotional level. A couple might go through the motions of an amicable relationship even though the guilt and resentment of a previous betrayal still simmer underneath. Previous enemies might form an alliance for expedience without really trusting each other. These partial forms of forgiveness make adaptive sense and have proximate mechanisms, but they fall short of the cathartic emotional experience in which a previous transgression is truly forgiven and the relationship continues as if it never happened. What people mean when they talk about "the joy of true forgiveness" in everyday language remains sensible within our more formal evolutionary framework, but it might be a relatively uncommon event. TFT and tinker-toy models notwithstanding, more sophisticated models will almost certainly confirm what already appears reasonable: The betrayal of a relationship is a very serious matter, in pairs or in larger groups, especially when the consequences are of a life-and-death magnitude. From the adaptive standpoint, a betrayal should be truly forgiven only when a change in the person or situation has made it truly irrelevant as far as future interactions are concerned.

THE ANTHROPOLOGY OF FORGIVENESS

So far I have shown that forgiveness has a biological foundation that extends throughout the animal kingdom. In an affiliated project, anthropologist Chris Boehm and I are studying forgiveness in traditional human societies around the world. It is actually difficult to find descriptions of forgiveness in hunter-gatherer societies, not because forgiveness is absent but because it happens so naturally that it often goes unnoticed. An exception is Turnbull's (1965) ethnography of the Mbuti (formally called Pygmies), which describes the life of a single band in novelistic detail. The following account of transgression and forgiveness is so evocative of life in small groups that I quote it in its entirely:

> We had all eaten in the evening and were sitting around our fires. I was with Kenge and a group of the bachelors, talking about something quite trivial, when all of a sudden there was a tremendous wailing and crying from Cephu's camp. A few seconds later there was shouting from the

path connecting the two camps and young Kelemoke came rushing through our camp, hotly pursued by other youths who were armed with spears and knives. Everyone ran into the huts and closed the little doorways. The bachelors, however, instead of going in with their families, ran to the nearest trees and climbed up into the lower branches. I followed Kenge up one of the smaller trees and sat among the ants, whose persistent biting I hardly noticed. What was going on below took all my attention.

Kelemoke tried to take refuge in a hut, but he was turned away with angry remarks, and a burning log was thrown after him. Masisi yelled at him to run to the forest. His pursuers were nearly on top of him when they all disappeared at the far end of the camp.

At this point three girls came running out of Cephu's camp, right into the middle of our clearing. They also carried knives—the little paring knives they used for cutting vines and for peeling and shredding roots. They were not only shouting curses against Kelemoke and his immediate family, but they were weeping, with tears streaming down their faces. When they did not find Kelemoke, one of them threw her knife into the ground and started beating herself with her fists, shouting over and over again, "He has killed me, he has killed me!" After a pause for breath she added, "I shall never be able to live again!" Kenge made a caustic comment on the logic of the statement, from the safety of his branch, and immediately the girls turned their attention to our tree and began to threaten us. They called us all manner of names, and then they fell on the ground and rolled over and over, beating themselves, tearing their hair, and wailing loudly.

Just then there came more shouts from two directions. One set was from the youths who had evidently found Kelemoke hiding just outside the camp. At this the girls leaped up and, brandishing their knives, set off to join the pursuit. The other shouts, for the first time from adult men and women, came from the camp of Cephu. I could not make out what they were about, but I could see the glow of flames.

I asked Kenge what had happened. He looked very grave now and said that it was the greatest shame that could befall a Pygmy—Kelemoke had committed incest. In some African tribes it is actually preferred that cousins should marry each other, but among the Ba-Mbuti this was considered almost as incestuous as sleeping with a brother or sister. I asked Kenge if they would kill Kelemoke if they found him, but Kenge said they would not find him.

"He has been driven to the forest," he said, "and he will have to live there alone. Nobody will accept him into their group after what he has done. And he will die, because one cannot live alone in the forest. The forest will kill him. And if it does not kill him he will die of leprosy." Then, in typical Pygmy fashion, he burst into smothered laughter, clapped his hands and said, "He has been doing it for months; he must have been very stupid to let himself be caught. No wonder they chased him into the forest." To Kenge, evidently, the greater crime was Kelemoke's stupidity in being found out.

All the doors to the huts remained tight shut. Njobo and Moke had both been called on by the youths to show themselves and settle the matter, but they had both refused to leave their huts, leaving Kelemoke to his fate. Now there was an even greater uproar over in Cephu's camp, and Kenge and I decided to climb down from the tree and see what was going on.

One of the huts was in flames, and people were standing all around, either crying our shouting. There was a lot of struggling going on among a small group of men, and women were brandishing fists in each other's faces. We decided to go back to the main camp, which by then was filled will clusters of men and women standing about discussing the affair. Not long afterward a contingent from Cephu's group came over. They swore at the children, who, delighted by the whole thing, were imitating the epic flight and pursuit of Kelemoke. The adults, in no joking frame of mind, sat down to have a discussion. But the talking did not center around Kelemoke's act so much as around the burning of the hut. Almost everyone seemed to dismiss the act of incest. One of Kelemoke's uncles, Masalito, with whom he had been staying since his father died, had great tears rolling down his cheeks. He said, "Kelemoke only did what any youth would do, and he has been caught and driven to the forest. The forest will kill him. That is finished. But my own brother has burned down my hut and I have nowhere to sleep. I shall be cold. And what if it rains? I shall die of cold and rain at the hands of my brother."

The brother, Aberi, instead of taking up the point that Kelemoke has dishonored his daughter, made a feeble protestation that he had been insulted. Kelemoke should have taken more care. And Masalito should have taught him better. This started the argument off on different lines, and both families quarreled bitterly, accusing each other for more than an hour. Then the elders began to yawn and say they were tired; they wanted to go to sleep; we would settle it the next day.

For a long time that night the camp was alive with whispered remarks, and not a few rude jokes were thrown about from one hut to another. The next day I went over to Cephu's camp and found the girl's mother busy helping to build Masalito's hut; the two men were sitting down and talking as though nothing much had happened. All the youths told me not to worry about Kelemoke, that they were secretly bringing him food in the forest, he was not far away.

Three days later, when the hunt returned in the late afternoon, Kelemoke came wandering idly into the camp behind them, as though he too had been hunting. He looked around cautiously, but nobody said a word or even looked at him. If they ignored him, at least they did not curse him. He came over to the bachelor's fire and sat down. For several minutes the conversation continued as though he were not there. I saw his face twitching, but he was too proud to speak first. Then a small child was sent over by her mother with a bowl of food, which she put in Kelemoke's hands and gave him a shy, friendly, smile.

Kelemoke never flirted with his cousin again, and now, five years later, he is happily married and has two fine children. He does not have leprosy, and he is one of the best liked and most respected of the hunters. (Turnbull 1965, 111–14)

Retaliation, a lasting change in behavior, and forgiveness took place through an outpouring of emotions. The group weathered the storm and sailed along on a relatively even keel without anyone knowing that they were at the helm. This episode recalls Tocqueville's description of small groups as so perfectly natural that they seem to constitute themselves.

Tocqueville appreciated that large societies are not natural in the same way. Something more must exist for a nation such as France or the United States to hang together. Furthermore, the "something" that in Tocqueville's time held France together was different from the "something" that held the United States together, with important consequences for the vitality of the two nations. The "somethings" were a complex mix of psychological attitudes, informal customs, and formal social organizations that can be collectively referred to as culture. Of course, hunter-gatherer societies such as Mbuti also have a culture that is important for orchestrating their behavior. We are talking about differences in culture and not its presence and absence. Rather than calling small societies "natural" and large societies "unnatural," we should simply say that they require different cultures to make them hang together as societies. The Mbuti do not have kings, courts, or constitutions. Larger nations do—and must—to exist at the scale that they do. Far from marginalizing culture, innate psychology provides the building blocks from which innumerable cultural structures have been built.

When human societies increased in scale beyond hunter-gatherer groups, forgiveness became culturally elaborated in ways that have attracted the notice of anthropologists. A good example is the Nuer, whose religion was described in chapter 2. Nuer society is segmented into a nested hierarchy of groups. The smallest groups are village size and function as corporate units (the term used by Evans-Pritchard) in the daily round of life, which includes tending their all-important cattle, growing millet, and providing for collective defense. The smallest groups often fight with each other but easily unite to form a corporate unit of their own when necessary, usually for cattle raiding or defense of their own cattle.

In the following passage, Evans-Pritchard describes the character of the Nuer people. It might seem strange and even politically incorrect to talk about people in a given culture as having a distinctive character. Do we not all share roughly the same genes and innate psychology? Perhaps, but bricks can be used to build many structures; from churches, to homes, to fortresses, to prisons, and the same can be said for the building blocks of culture. Evans-Pritchard writes:

> The Nuer is a product of hard and egalitarian upbringing, is deeply democratic, and is easily roused to violence. His turbulent spirit finds any restraint irksome and no man recognizes a superior. Wealth makes no difference. A man with many cattle is envied, but not treated differently from a man with few cattle. Birth makes no difference. A man may not be a member of the dominant clan of his tribe, he may even be of a Dinka descent, but were another to allude to the fact he would run a grave risk of being clubbed. (1940, 181)

The fierce egalitarianism that holds hunter-gatherer society together also provides the foundation for Nuer society. And egalitarianism is enforced with violence. The cardinal virtue for a Nuer male is to defend his honor and the honor of his group, which is defined with great flexibility according to the situation (recall the concept of trait-groups discussed in chapter 1). Retaliation truly is divine among the Nuer. However, this hair-trigger capacity for violence requires a counterbalance which exists in the form of a person called the leopard-skin chief, who plays a vital role in nonlethal conflict resolution. When defense of honor leads to a homicide, the killer can receive sanctuary in the home of the leopard-skin chief. Over a period of weeks and months, the leopard-skin chief negotiates a settlement in which, rather than a revenge killing, cattle are paid for the slain man's life. The negotiations are delicate because no individual or group wants to gain a reputation for failing to defend their honor to the fullest degree. Nevertheless, behind the braggadocio is a willingness to negotiate, as long as there are reasons for the two sides of the conflict to get on with the daily round of life. Homicides that take place within a corporate unit are patched up relatively easily, but a killer from a distant group may never be truly forgiven, even after the payment of cattle. If the killer is from another tribe, there is no question of cattle payment and violent retaliation is the only solution.

So far I have described the leopard-skin chief and his function in Nuer society in purely secular terms. However, the leopard-skin chief is also a sacred personage and must be so to perform his secular function. He has no real power and is even chosen from an unimportant lineage so that he can be nonpartisan when negotiating with the more powerful lineages. There is nothing whatsoever preventing the victim's kin from rushing into the leopard-skin chief's compound to kill the slayer, other than respect for the sacred office. Evans-Pritchard describes the leopard-skin chief as "a sacred person without political authority" (1940, 5). Recall from chapter 4 that the Balinese combined a hard-headed knowledge of the pragmatic side of their religion with "overflowing belief" in its sacred dimension. Similarly, the Nuer could explain the function of the leopard-skin chief in completely pragmatic terms—"We took hold of them and gave them leopard skins and made them our chiefs to do the talking at sacrifices for homicide" (173)—and yet sincerely pay them the respect embodied in the word "sacred." Evans-Pritchard may have disagreed with Durkheim at some level, but this is surely an example of what Durkheim meant when he said "social life is only possible thanks to a vast symbolism."

It is hard to tell whether the Nuer ever feel "the joy of true forgiveness" in their attempt to resolve conflicts. However, Boehm's (1984, 133–34) ethnography of Montenegro, one of the last tribal societies of Europe, includes a description of a forgiveness ceremony in the voice of one of the participants, as spoken to Mary Durham, a "spirited Englishwoman" who wrote a book about the Balkans based on her travels in the 1920s (Durham 1928). I quote the narrative in full because it illuminates a number of themes that we are trying to incorporate into our theoretical framework. Parenthetical statements to clarify the narrative were added by Durham, and I have added a few of my own in brackets.

"In a few days the village gathered and wanted us to make peace. We sent men to them and asked for the first truce till St. Dmitri's Day (October 26th); and so soon as it came we asked for a second till Christmas, which they granted after much begging. At Christmas we asked, as is the custom, for both truce and arbitration ('kmetstvo')." (If a third truce were granted this meant that arbitration would be accepted. If a third truce were refused the feud raged as badly as ever.) "We fixed it for St. Sava's Day (January 14th). They gave us the names of twenty-four men [respected members of the community that collectively arbitrate the proceedings], and off went I over wood and rock to beg them to come, and luckily none refused. St. Sava's Day came. I killed two oxen and six sheep, took four hams and bought two barrels of wine. I gathered together my bratstvo, my Kums, and my pobratims [allies in the conflict , including both genetic kin and 'sworn brothers'], and, God forgive me, they helped me with money and bread, and so I had all that was needed. And the men sat down and gave judgement thus. They held the head of Nikola Perovo as equivalent to that of my dead father. The head of Gjuro Trpkov they valued at 120 zecchins. One of their wounded was held equivalent to my wounds and the other was valued at seven bloods" (one "blood" was 10 zecchins; . . . the judges valued the wounds by this standard), "and that woman's wound was reckoned as three bloods. And they decreed that I should bring six infants" (in order that a man of the other bratstvo shall stand godfather to them and thus cement peace by a spiritual relationship), "and that I should hang the gun which fired the fatal shot around my neck and go on all fours for forty or fifty paces to the brother of the deceased Nikola Perova. I hung the gun to my neck and began to crawl towards him, crying: 'Take it, O Kum, in the name of God and St. John.' I had not gone ten paces when all the people jumped up and took off their caps and cried out as I did. And by God, though I had killed his brother, my humiliation horrified him, and his face flamed when so many people held their caps in their hands. He ran up and took the gun from my neck. He took me by my pigtail ('perchin') and raised me to my feet, and as he kissed me the tears ran down his face, and he said: 'Happy be our Kumstvo (Godfatherhood).' And when we had kissed I, too, wept and said: 'May our friends rejoice and our foes envy us.' And all the people thanked him. Then our married women carried up the six infants, and he kissed each of the six who were to be christened."

"Then all came to us and sat down to a full table." (They are waited on by the head of the house and his men, who do not sit down with the elders and the plaintiff. At the head of the table sits the most respected of the elders. After the meal he proposes the health of the new Kum and of the master of the house, and they drink to the newly established peace. The payment of the fine is then called for.) " 'By God, my brother,' said my uncle, 'we have but little money. But we are a fine bunch of brethren, each with shining weapons. Here they are and here are you. Another time we shall do to you as you do to us. Here are the weapons. Take what your honour permits you.' Kum Kikola was indeed a man. He took the gun which had shot his brother and kissed it on the muzzle. The other weapons he gave back, saying: 'Take them, O Kum. I give them you as gift in return for the six Godfatherhoods; and I give you my brother for the gun.' " (Durham 1928, 89–90)

This narrative is laced with Christian imagery, suggesting that the Christian religion per se was an important element of the forgiveness ceremony. However, we should reserve judgment on this issue. Christianity is well known to accommodate itself to the cultures of the people it "converts." Montenegrin priests performed important functions, but they made war like everyone else and were not even mentioned in the narrative quoted above. The essential elements of the forgiveness ceremony can be described without reference to Christianity: the will of the community, arbitration by members of the community that are respected by both sides of the conflict, an expensive commitment of resources by the side that is currently ahead in the death count, a deliberate stitching together of the two sides through the device of godparenthood, which obligates the adults on one side to care for the infants of the other side, and a rite of humiliation that may have been the most difficult part of all, given the proud nature of the Montenegrin people. There can be no doubt that this cultural symphony accomplished a psychological transformation that can be accurately described as "the joy of true forgiveness," although we are not told how long it lasted.

The narrative also hints at a larger cultural symphony that orchestrates the lives of the Montenegrin people, quite apart from this particular ceremony. Most people regard feuding as a social pathology, but it is properly understood as part of an intensely moral system that keeps the peace much more than breaking it. Boehm (1984, 187–88) makes a telling analogy with automobile fatalities in America:

If the reader is still unconvinced that an intelligent group of people could knowingly sustain such a "menace" in their own midst, it might be useful to think of the way in which we Americans handle our own "automobile menace" as an ecological problem. These vehicles kill approximately fifty thousand Americans each year, or one person in every five thousand annually, and they seriously injure many more. . . . These losses far outstrip our total losses in all wars. A question arises, since we are a scientifically minded people, supposedly committed to enlightened public policy, and are possessed of a centralized government that has fairly strong coercive sanctions: How is it that we have allowed this to take place as casually as we have?

I do not believe that the answer is that "democracy does not work" or that no one has really added up the entire problem or—certainly—that we recognize automotive fatalities as a needed means of population control. We are conscious both of the high human costs of running automobiles and of the much lower cost (in human lives) of virtually any other form of transportation. Yet we are also aware, perhaps more on an intuitive level, of the gains that we derive from our beloved cars. They provide us with the freedom to go where we please and to live where we please; and for many they serve as a basic expression of aesthetic taste, of social status, and, frequently, of personal potency in general. In short, for Americans the "killer-automobile" is closely tied to many of the things that we value most in life. We do regulate its potential destructive power by creating and sanctioning rules that apply to its use, but we never consider outlawing it or seriously restricting its ownership and use. And while the damage is considerable, at least it is fairly predictable. If we are prepared to accept the losses, then we can go about life believing that we have the situation under control, which we do.

It turns out that the mortality rate from feuding in the Balkans is about one half the mortality rate from automobiles in America (although the feuding rate was probably greater during past centuries). Even after we appreciate Boehm's point, however, we must keep a balanced perspective. Feuding in tribal societies and automobile fatalities in modern societies definitely belong on the cost side of the ledger. Perhaps they are linked to even larger benefits, but they may not be. To describe a culture as a symphony or a house does not mean that every note is perfect or that every brick is in place. When the symphony is written and the house is built by an evolutionary process, they assuredly will not be perfect. I can easily imagine a social movement such as Mothers Against Drunk Driving (MADD) permanently reducing traffic fatalities by implementing laws and changing the "national consciousness"—a term that is perfectly acceptable from the perspective of multilevel selection theory. In addition, even highly adaptive structures often cannot easily evolve into radically different structures. A mouse cannot easily evolve into a giraffe, a church cannot easily be renovated into a prison, and it might be very difficult to turn a feuding society into a modern democracy, no matter how much it would profit from the transition (Putnam 1992; Fukuyama 1995). A sophisticated understanding of evolution recognizes both the presence and absence of functional design. However, it is the presence of functional design at the level of whole cultures that needs to be emphasized against the background of the last five decades of intellectual thought in biology and the social sciences.

We can summarize the progress that we have made in this chapter as follows: Many attributes that are often regarded as uniquely human are in fact biologically ancient because they are adaptive and don't require that much brain power to implement. However, this does not even remotely diminish the importance of culture in human affairs. Cultures are required to orchestrate human behavior in relation to specific environments. An infinitude of cultures are possible to channel our innate psychological impulses in different directions. Most possible cultures are not adaptive, and we should hope for our own sakes that a process exists for winnowing among the many possibilities, leaving a subset that are at least somewhat well-adapted to their environments. The larger human groups become, the more culture is required to channel the emotional outpourings of our innate psychology, which was originally designed to work primarily in small groups. Forgiveness—the specific subject of this chapter—needs to be seen against this background as a highly orchestrated and context-sensitive behavior whose expression can vary adaptively among cultures.

Christian Forgiveness

A religion such as Christianity must orchestrate the behavior of its members, much like the traditional cultures described in the previous section. This is a daunting task when we consider its magnitude. Many activities must take place among many different categories of people. In-group and out-group relationships require different sets of instructions. Out-group relationships might themselves need to be subdivided into radically different types. The most appropriate activities are likely to change with the physical and social environment. Moreover, all of this information must be transmitted to converts who enter the faith with very different beliefs and practices. Even without knowing the details, we can be certain that the instructions for behaving adaptively, which somehow are encoded in the beliefs and practices of the religion, must be complex.

The early Christian Church provided a code that was highly adaptive for its members, as we learned in chapter 4. If a Christian was asked to summarize her religion in a single word, it might be "forgiveness." But I have just shown that an adaptive religion cannot be summarized in a single word, other than "complex." To understand the nature of Christian forgiveness, we must study it as part of a more comprehensive system of if-then rules that functions in the context of a particular environment.

Fortunately, the environment of the early Christian Church is understood in considerable detail. The following account draws upon Pagels (1995), who represents a branch of religious scholarship somewhat separated from the rational choice literature. Her account of Judaism in the ancient world closely matches the second hypothesis for the life cycle of religions outlined in chapter 5. The religious establishment was constantly in danger of being "corrupted" by those who profited at the expense of others within their own faith and who were tempted to abandon the strict laws that kept Israel separated from the other nations—to become more church-like, in the terminology of Stark and Bainbridge. This trend was opposed by passionate efforts to "purify" the church—to make it more sect-like—led mostly by those who felt exploited by their own upper class.

This tension often led to a power struggle for leadership of the main religion. If the purifying faction is sufficiently strong, the same dynamic that gives rise to derivative sects can also move the entire religion in a sect-like direction. The Maccabean revolt provides one example:

> As told in 1 Maccabees, this famous story shows how those Israelites determined to resist the foreign king's orders and retain their ancestral traditions battled on two fronts at once—not only against the foreign occupiers, but against those Jews who inclined toward accommodation with the foreigners, and toward assimilation. Recently the historian Victor Tcherikover and others have told a more complex version of that history. According to Tcherikover, many Jews, especially among the upper classes, actually favored Antiochus's "reform" and wanted to participate fully in the privileges of Hellenistic society available only to Greek citizens. By giving up their tribal ways and gaining for Jerusalem the prerogatives of a Greek city, they would win the right to govern the city themselves, to strike their own coins, and to increase commerce with a worldwide net-

work of other Greek cities. . . . . They could advance themselves in the wider cosmopolitan world.

But many other Jews, perhaps the majority of the population of Jerusalem and the countryside—tradespeople, artisans, and farmers—detested these "Hellenizing Jews" as traitors to God and Israel alike. The revolt ignited by old Mattathias encouraged people to resist Antiochus's orders, even at the risk of death, and oust the foreign rulers. After intense fighting, the Jewish armies finally won a decisive victory. They celebrated by purifying and rededicating the Temple in a ceremony commemorated, ever since, at the annual festival of Hanukkah.

Jews resumed control of the Temple, the priesthood, and the government; but after the foreigners had retreated, internal conflicts remained, especially over who would control these institutions. These divisions now intensified, as the more rigorously separatist party dominated by the Maccabees opposed the Hellenizing party. The former, having won the war, had the upper hand. (Pagels 1995, 45–46)

At other times, the worldly faction held sway, and the purifying faction left the main body of Judaism to form sects of their own that they regarded as the true nation of Israel. The Essenes were such a sect. The sect founded by Jesus was another, as his many diatribes against the Jewish priests and scribes and his clearing of the Temple makes clear.

The Jesus movement was similar to other Jewish sects in some respects but fundamentally different in others. All of the sects were struggling against the main body of Judaism that constituted the most severe threat to their existence, compared to the more remote threat posed by the Roman Empire or other nations. For all sects, the mere fact of being a Jew no longer sufficed to be counted among the chosen people. One must also uphold the laws of the covenant, as defined by the sect. It is here that Jesus parted company with other sects such as the Essenes. For the Essenes, the laws of the covenant were an ultra-strict form of Judaism that could be practiced only by Jews. The mere fact of being a Jew was not sufficient, but it was assuredly necessary to be counted among the chosen. Jesus emphasized the moral side of Jewish law more than its ethnic or ritual side, so much that the mere fact of being a Jew ultimately became neither necessary nor sufficient to be among the chosen. This one structural change constituted an important point of difference between a tiny extinct sect known only through the Dead Sea Scrolls and a religious movement that spread around the entire globe.

So far I have recounted a history of the origin of the Christian Church, similar in style to the history of life on earth. How did mammals originate and why did they come to replace the dinosaurs as the dominant vertebrate life form on our planet? We can never be certain, but even if we could, the story would be full of happenstance and serendipity. As Stephen Jay Gould (1989) has pointed out, if we could replay the tape of life it would not necessarily turn out the same. Mammals could have remained insignificant mouse-sized creatures, and human-like intelligence could still be absent on the planet. Similarly, the sect founded by Jesus can be regarded as a mutated culture whose structural properties can be pinpointed with impressive detail. The fact that it originated when and where it did is partly a matter of happenstance. History could have taken a different course. However, given the course that history did take, chance started to play a much smaller role. As Pagels notes, Jesus and his disciples may never have anticipated or even desired their new sect to attract so many gentiles, but there was little to be done about it now that the mutated culture had come into existence.

Now that we understand the social environment, we can attempt to predict the adaptive properties of the early Christian Church in more detail. At least four major categories of outsiders can be distinguished; the worldly Jewish establishment that posed the greatest threat, the more remote but still threatening Roman Empire, potential Jewish converts, and potential gentile converts. To survive and prosper, the early Christian Church must have provided mechanisms for isolating its members from the outside world, for motivating their behavior as strongly as possible, for orchestrating their behavior toward each other, and for orchestrating their behavior toward the four categories of outsiders. Potentially, the rules governing forgiveness might need to be different for each of these relationships.

The Four Gospels are presented as historical accounts of the life of Jesus. Voluminous religious scholarship has determined that they were written between thirty-five and one-hundred years after his death. The differences between the Four Gospels are sometimes explained simply as the fading and embellishment of memory, in which case the earliest account (Mark) should be most accurate and the latest account (John) most distorted. Pagels and many of her colleagues reject this interpretation in favor of one based on utility. According to Pagels, all four Gospels function as how-to manuals enabling local congregations to function as adaptive units. The instructions are encoded in the historical narratives, which makes all four Gospels suspect as literal history. Narratives designed to motivate behavior are free to omit, distort, and make up facts whenever necessary. The Four Gospels differ from each other, not because they were separated in time, but because they were designed to serve the needs of different Christian churches scattered across the Roman Empire. They are a fossil record of cultural adaptation at an extremely fine scale. Just as upstream and downstream guppy populations evolved to be different in response to the presence and absence of predators, the instructions provided by the four Gospels evolved to be different in response to differences in the social environments inhabited by the early Christian congregations.

Let us see how stories, including the life of Jesus and the parables that he told, provide a set of if-then rules that regulate forgiveness in addition to many other forms of behavior. The first requirement for organizing a group into an adaptive unit is to define the group and isolate it from the rest of society so that in-group and out-group behaviors can be regulated separately. Of course, anyone familiar with the Gospels knows that Jesus demanded total commitment, eclipsing not only one's prior religion but also one's immediate family. As one of many examples, consider the well-known parable of the talents in Luke 19:12–27. Jesus tells a story of a nobleman who travels to a distant land "to get royal power for himself and then return." People who have attended Sunday School or Bible study know all about the servants, wise and foolish, who are entrusted with their master's resources while he was away. But what people may not notice is that, upon his return, the nobleman demands the death of those who were disloyal in his absence: "As for those enemies of mine who did not want me to be king over them—bring them here and slaughter them in my presence." This facet of Christianity is the polar opposite of forgiveness, and must be so, to perform its basic function. Before there can be a strongly committed group, there must be perceived dire consequences for leaving the group.

Another requirement is to motivate the behavior of group members. Fighting for your life is more motivating than fighting for a seat on the bus. Fighting the final cosmic battle of good against evil is more motivating than fighting for your life. The Gospels pull out all the stops in their motivation of behavior, as only a religious belief system can. According to Pagels, the Essenes and Christians were notable for perceiving their own conflicts in terms of a battle between God and Satan. The impact of this belief upon behavior is vividly illustrated by the true story of Justin, a young man who traveled from Asia Minor to Rome in about 140 C.E. to pursue his passionate interest in philosophy. One day he went with his friends to the amphitheater to see the spectacular life-and-death gladiatorial fights. Pagels takes up the story from here:

Justin was startled to see in the midst of this violent entertainment a group of criminals being led out to be torn apart by wild beasts. The serene courage with which they met their brutal public execution astonished him, especially when he learned that these were illiterate people, Christians, whom the Roman senator Tacitus had called "a class of people hated for their superstitions," whose founder, Christos, had himself "suffered the extreme penalty under Pontius Pilate" about a hundred years before. Justin was profoundly shaken, for he saw a group of uneducated people actually accomplishing what Plato and Zeno regarded as the greatest achievement of a philosopher—accepting death with equanimity, an accomplishment which the gladiators' bravado merely parodied. As he watched, Justin realized that he was witnessing something quite beyond nature, a miracle; somehow these people had tapped into a great, unknown source of power.

Justin would have been even more startled had he known that these Christians saw themselves not as philosophers but as combatants in a cosmic struggle, God's warriors against Satan. As Justin learned later, their amazing confidence derived from the conviction that their own agony and death actually were hastening God's victory over the forces of evil, forces embodied in the Roman magistrate who had sentenced them, and, for that matter, in spectators like Justin himself. (1995, 115)

Seeing Christians die with composure actually caused this earnest man to become a Christian and literally rush to his own martyrdom. Although Justin is an extreme example, it appears to be true in general that persecution often increases the recruitment of new members to a sect by providing a form of honest advertisement (e.g., Quakerism; Ingle 1994). The fact that believers keep their faith despite the cost of persecution proves that they are gaining something precious from their religion. It therefore can make sense to join a persecuted sect and work hard to end the persecution, thereby gaining the full benefits without the cost. In any case, such extreme motivation does not always result in forgiveness. Satan, and those in league with him, are not forgiven. The Christians who were torn apart by wild beasts did not forgive their adversaries but rather stood up to them with unyielding courage, which in their powerless position took the form of death with composure. In the more powerful position that Christians imagined for themselves as warriors in the decisive battle between God and Satan, the same judgmental attitude expressed itself as the eternal damnation of the enemy. There is nothing forgiving about Christianity in this context.

An adaptive religion that is not serving an already defined ethnic group must obviously attract members. Like most sects, the early Christian Church was designed to benefit those who had become neglected and abused by worldly society. Its intense egalitarianism and commitment to altruistic love also attracted wealthy members, such as Justin, who cherished these values regardless of their own station in life. It is in this context that Christianity shows one of its forgiving sides. Anyone could enter the Church, no matter how poor, wretched, or even despicable. The fact that outcasts such as prostitutes and gentiles (from the perspective of Judaism) are prominently included among Jesus' followers vividly illustrates this facet of Christianity.

Although almost any past sin could be forgiven, it would be absurd to suppose that converts were free to continue their old ways after joining the church. One of the miraculous aspects of Christianity was the transformation that took place upon baptism. Here again is Pagels's account of Justin:

> Seeking out other "friends of Christ," Justin asked to become a candidate for the rite of baptism. He does not tell us the story of his own baptism, but other sources suggest the following: Having fasted and prayed to prepare himself, Justin would await, probably on the night before Easter, the rite that would expel the indwelling demonic powers and charge him with new, divine life. First the celebrant would demand to know whether Justin was willing to "renounce the devil, and all his pomp, and his angels"; Justin would ritually declare three times, "I renounce them." Then Justin would descend naked into a river, immersing himself to signify the death of the old self and the washing away of sins. Once the divine name was pronounced and the celebrant had invoked the spirit to descend on him, he would emerge reborn, to be clothed with new white garments at the shore and offered a mixture of milk and honey—babies' food, suitable for a newborn.
>
> Justin said that he had received in baptism what he had sought in vain in philosophy: "this washing we call illumination; because those who learn these things become illuminated in their understanding." He later explained to other potential converts, "Since in our birth we were born without consciousness or choice, by our parents' intercourse, and were brought up in bad habits and evil customs," we are baptized "so that we may no longer be children of necessity and ignorance, but become the children of choice and knowledge." His ritual rebirth to new parents—God and the holy spirit—enabled Justin to renounce not only

his natural family but the "habits and evil customs" they had taught him from childhood—above all, traditional piety toward the gods, whom he now saw as evil spirits. (1995, 118–19)

We can well imagine this ritual, coupled with a sincere belief in spirits, having a transformational effect on those who were born again.

Church members were expected to practice extreme altruism toward each other, which was illustrated again and again in the life of Jesus and the parables that he told. Of course, Jesus is the ultimate positive role model for altruism. It would be naive to think that such a lofty ideal could be approached without social control. Thus, forgiveness takes on yet another face as far as within-group interactions are concerned. The Gospels include key figures that deliberately betray Jesus (Judas) or wilt under pressure (Peter). Judas is not forgiven by Jesus, at least according to Mark 14: 18–21: "For the Son of Man goes as it is written of him, but woe to that one by whom the Son of Man is betrayed! It would have been better for that one not to have been born." Peter is forgiven but only after repenting for his previous denial of Jesus. According to Pagels, these narratives are constructed to have a powerful effect on the audience for whom the Gospels were originally written:

> Mark knows that those who publicly confess their conviction that Jesus is "the Messiah, the Son of God" (14:61) may put themselves in danger of abuse, ridicule, even threats to their lives. The terms Messiah and Son of God would probably have been anachronistic during the time of Jesus; but many of Mark's contemporaries must have recognized them as the way Christians of their own time confessed their faith. In this dramatic scene, then [Peter's denial of Jesus], Mark again confronts his audience with the question that pervades his entire narrative: Who recognizes the spirit in Jesus as divine, and who does not? Who stands on God's side, and who on Satan's? By contrasting Jesus' courageous confession with Peter's denial, Mark draws a dramatic picture of the choice confronting Jesus' followers: they must take sides in a war that allows no neutral ground. (28)

Religion as it is practiced can only hope to approximate religion as it is idealized. Jesus did not live long enough to deal with the transgressions that inevitably occur in the life of any community. However, his disciples were faced with transgressions on a daily basis and described a policy of measured forgiveness in their letters that were also included in the New

Testament. As I have already described for Calvinism in chapter 3, transgressions were punished in an escalating fashion resulting ultimately in
exclusion. It was Paul who said: "Cast out the wicked person from among
you" (1 Cor. 5:13). In all cases the goal was to eliminate the offensive
behavior, not the person. Thus, even the most grievous sin committed by
a church member could be forgiven with sufficient contrition, as judged
by the church. Belief in spirits probably facilitated the distinction between
a person and the person's behavior. A person who committed evil acts
was not necessarily an evil person but rather inhabited by an evil spirit.
By expelling the spirit the person can be truly forgiven, in the psychological sense of the term, because the person was just a vessel, not at fault.
Thus, the extreme forgiveness that allowed anyone to enter the church was
also extended to church members, but not in a way that allowed exploitation to occur within the church. This kind of conditional forgiveness is
required for any human group to function as an adaptive unit.

To summarize our progress so far, the Gospels provide a code for creating and isolating the church from its surrounding society, powerfully motivating its members, and instructing them how to behave among themselves and toward outsiders. Forgiveness has many faces—*and needs to*—
in order to function adaptively in so many different contexts. However,
the category of outsider has not yet been sufficiently differentiated. There
is a policy of extreme altruism and forgiveness toward the downtrodden
(even if they are not likely recruits), and a policy of unyielding opposition
toward those who are in league with Satan. But who is in league with
Satan? For Jesus and the authors of the Gospels, the main Jewish religious
institutions were the most important enemy, not the Roman Empire. This
adversarial relationship, and corresponding lack of forgiveness, are reflected in the historical account of Christ's death reported in the Gospels.

Little is known about the historical figure of Jesus, but more is known
about Pontius Pilate, the Roman governor of Judea at the time of Christ's
death. By all accounts he was a cruel man accustomed to using brutal
tactics to get his way. According to Pagels (29–30), contemporary accounts
range from bitterly hostile to negative. Philo, an influential member of the
Jewish community in Alexandria, described Pilate as a man of "inflexible,
stubborn, and cruel disposition" whose administration was marked by
"greed, violence, robbery, assault, abusive behavior, frequent executions
without trial, and endless savage ferocity." In the Gospels, however, he is
variously described as a weak person who caves in to the demands of the
Jews or as a just person who proclaims the innocence of Jesus no less than
three times before the Jews finally get their way. In short, Pontius Pilate

is presented in a way that shifts the responsibility for the death of Jesus away from the Romans and toward the Jews.

So far I have discussed elements of the Gospels that are largely shared by all four versions. In addition, Pagels interprets the differences among the Gospels as adaptive responses to the local pressures faced by the congregations for whom each Gospel was written, with corresponding implications for the rules of forgiveness. The Gospel according to Mark was written in the immediate aftermath of the Roman siege of Jerusalem and destruction of the Temple in 70 C.E. The Romans were no friends of the Christian Church for whom the Gospel was written, but the establishment Jews were still the primary enemy and disenfranchised Jews the primary source of converts. In fact, the destruction of the Temple was interpreted as God's punishment for the sins of the establishment Jews, foretold by Christ, and a sure sign of his imminent return. As the primary adversary, the Jewish establishment is portrayed as primarily responsible for the death of Jesus.

For Matthew, writing only ten to twenty years later, the main body of Judaism was controlled largely by a religious party known as the Pharisees, which was far less powerful during the time of Jesus. Nevertheless, the Pharisees became the primary targets for blame in the Gospel according to Matthew. Luke was apparently a gentile convert writing for a gentile Christian Church. John was probably a member of a radically sectarian church composed of Jews in even more bitter opposition to the establishment than usual. Since the main purpose of each Gospel was to guide the behavior of a particular church, each became locally adapted to a particular social environment.

Pagels develops this thesis in impressive detail, but I will confine myself to the Gospel according to Luke, which was probably written for a gentile audience before the Christian movement had grown large enough to attract the attention of the Roman establishment. In the other Gospels, Jesus' first episode of public teaching in his home town of Nazareth leads to a favorable reception, but in Luke it leads to a riot because Jesus predicts that God will bring salvation to the gentiles, even at the cost of bypassing the Jews. At this news, "all in the synagogue were filled with rage. They got up, drove him out of the town, and led him to the brow of the hill on which their town was built, so that they might hurl him off the cliff" (Luke 4:28–29).

This passage shows that Luke had little reason to temper his description of the Jews or to distinguish the wealthy establishment from the rank and file. He did, however, have a good reason to cast the Romans in a favorable light. According to Luke, Pilate does his level best to save Jesus from the Jews. First he proclaims "I find no basis for an accusation against this man" (23:4). Then he tries to wash his hands of the responsibility by turning Jesus over to Herod, the Roman-appointed King of Judea, for judgment. It is Herod's Jewish officers who beat and mock Jesus, not the Roman soldiers. When Herod sends Jesus back and Pilate again must decide, he eloquently defends Jesus against both the Jewish leaders *and* the people, who cry out in one voice for his death. Even then Pilate defends Jesus for a third time, saying "Why, what evil has he done? I have found in him no ground for the sentence of death; I will therefore have him flogged and then release him" (23:22). But he must ultimately give Jesus over "to their will." It is left unclear who actually crucifies Jesus—Roman soldiers or a Jewish mob—but a Roman centurion who sees him die "praised God," and concludes "Certainly this man was innocent!" (23:47).

There seems to be no end to the tailoring of the Gospels to meet the utilitarian needs of the first local congregations, not only with respect to forgiveness but in general. The Virgin birth, which appears for the first time in Matthew, was probably a defensive reaction to the criticism that Jesus could not be the Messiah because he did not have a distinguished pedigree. Well, he does now! In the Gospel according to Matthew, Herod hears of Christ's birth and jealously kills all the male children under the age of two in the region around Bethlehem, a detail that is left out of the other Gospels and for which there is no independent evidence. How do we know that the disciples didn't steal Christ's body and fake the resurrection? Because there was a Roman soldier guarding that cave!

Even more interesting are the Gospels that didn't make it into the New Testament. When the Christian movement gained sufficient momentum, it became necessary to impose uniformity by canonizing some of the religious teachings and denouncing the rest. During the late second century, bishops of the Christian church that by now was calling itself orthodox met to assemble the New Testament and brand everything else, in the words of one of the bishops, "an abyss of madness, and blasphemy against Christ" (Pagels 1995, 70). Among the gospels pushed into the abyss were the Gospel of Thomas, which encourages the believer to embark upon a journey of self-discovery rather than conforming to a close-knit group. According to Pagels (74–75), the Gospels that made it into the New Testament were chosen for the following reason: "The author of Mark, then, offers a rudimentary model for Christian community life. The gospels that the majority of Christians adopted in common all follow, to some extent, Mark's example. Successive generations found in the New Testament gospels what they did not find in many other elements of early Jesus tradition—a practical design of Christian communities." Pagels's emphasis on practical design clearly dovetails with Durkheim's emphasis on secular utility and the central thesis of this book. In addition, even though Pagels does not use the word evolution, it is clear that she is describing an evolutionary process in which versions of Christian belief that work well at the community level replace those that don't. Forgiveness has many faces in the Christian beliefs that survived, as it must for Christians to behave adaptively in their complex social environments.

THE NONADAPTIVE SIDE OF THE EVOLUTIONARY COIN

So far I have emphasized the adaptive features of the early Christian Church at the behavioral level (prediction) and also at the level of proximate mechanisms (production). However, Pagels provides such a detailed analysis of churches in relation to their environments that we can also begin to see the nonadaptive side of the evolutionary coin. Recall that not everything is adaptive in nature, as evolutionary biologists such as Stephen Jay Gould (1989; Gould and Lewontin 1979) have emphasized. Even in individual organisms, proximate mechanisms can look more like Rube Goldberg devices than sleekly designed machines, and they can be downright dysfunctional in some of their manifestations. Despite its impressive adaptedness, in some respects the early Christian Church has a decidedly Rube Goldberg feel to it, with additional implications for the Christian concept of forgiveness.

One remarkable fact is that the gospel writers so freely altered their most sacred story to adapt it to their particular needs. When it comes to altering a sacred story, it seems that nothing is sacred—at least during the early stages of religious evolution. Another remarkable fact is that the Four Gospels could be assembled into a single collection in spite of their massive contradictions. Even a child approaching the New Testament with a clear eye must wonder why Jesus preached successfully in his home town according to one version and was almost thrown off a cliff in another. There must be a sense in which religious belief is so concerned with using sacred symbols for utilitarian purposes that logical consistency and historical accuracy become secondary considerations. Despite its internal contradictions, the New Testament provides an arsenal of sacred symbols that can be selectively employed depending on the specific situation. As one example, Pagels (99) notes that the Gospel according to John, whose sect was most bitterly opposed to its surrounding social environment, has ever since been an inspiration and comfort to Christian churches that find themselves battling for their lives.

The very success of the Christian Church may also have limited the evolutionary flexibility that characterized its early stages. Once the New Testament was canonized, the only way to adapt to future environments was to select from an unchanging arsenal of sacred symbols. It was impossible to make the Romans the primary culprit for Jesus' death, even though it would have made sense after the Romans became the primary enemy of the Church. In addition, it is entirely possible that the New Testament has predisposed Christians to hate Jews long, long after it ceased to be adaptive. In short, to the extent that Christian belief systems are not adapted to their current environments, they can cause inappropriate patterns of forgiveness and failure to forgive.

PILGRIM'S PROGRESS

The goal of this chapter is to understand the concept of forgiveness, often regarded as a hallmark of the Christian religion, from a multilevel evolutionary perspective. Thinking of churches as organisms encourages us to look for adaptive complexity, and in the case of forgiveness it has encouraged us to go beyond the simple injunction to turn the other cheek. I began by describing forgiveness as part of a profile of traits that is widely distributed throughout the animal kingdom because it is adaptive and doesn't require that much brain power. Far from marginalizing the concept of culture, this innate psychology provides the "musical instruments," or "building blocks," from which innumerable cultural "symphonies" or "structures" can be built. The larger human societies become, the more culture is required to adaptively regulate and coordinate human behavior. The first social institutions to appear when human societies increase in scale above hunter-gatherer groups are conflict resolution devices, such as the leopard-skin chief among the Nuer, without which the society would almost certainly fragment into smaller units. Furthermore, these institutions tend to be religious in Durkheim's sense of obtaining their power from a sense of sacredness. A religion such as Christianity must provide close to the entire culture for its people, a daunting task when we consider what is required to define a group, motivate its members, and define their relationship with each other and with numerous categories of outsiders. Not only does the Christian religion meet this challenge, but it even adjusts itself to the specific demands of local environments. This miracle of cultural adaptation is accomplished largely through powerful stories thinly disguised as historical narratives.

Unfortunately, whether we regard the Christian religion as a miracle depends upon our perspective. A believer might regard it as a miracle for different reasons than I do. An unbeliever might ridicule it for the same reasons that I call it a miracle. Let us try to unpack these complications caused by perspective, one by one.

First, we should acknowledge that we have achieved a satisfying account of Christian forgiveness as a complex adaptation that functions in a variety of different contexts. Context-sensitivity is the key for understanding the nature of Christian forgiveness. Ignoring context can only turn sense into nonsense. How can Christians preach forgiveness when they are so judgmental about people's behavior? How can they preach the Golden Rule while waging war? Should a true Christian forgive someone who commits a horrific crime that remains unpunished? One form of hell for me would be to be locked for eternity in a room full of people who make these observations, again and again, as if for the first time. These observations portray religion as stupid when in fact it is the observer who is stupid for failing to grasp the full complexity of what religion must do to serve its people. Much of what has been written about religion becomes obsolete as soon as we appreciate its complex adaptive nature.

Having made this point, we must acknowledge that the exalted view of Christianity must also be tempered. Christianity and virtually all other religions fall short when judged by the loftiest standard of universal brotherhood. They merely adapt groups to their local environments. When they lift people out of poverty and desperation, they deserve our highest admiration. When they become agents of conquest and aggression, they remain exalted for their believers but deserve to be judged as immoral by outsiders who see the aggressors as part of a larger group. Perhaps universal brotherhood can be achieved by a religion or another social organization, but that is a challenge for the future.

For me, the failure of religion to achieve universal brotherhood is like the failure of birds to break the sound barrier. Imagine that you discover a bird with an injured wing. By fixing its wing, you enable the bird to fly faster. Now imagine that you discover a bird that is perfectly healthy. It has been beautifully designed by natural selection to speed through the air at seventy miles per hour. However, you want to help it fly faster. Your task in this case will be much harder than fixing an injured wing. You will need to discover a design breakthrough that was missed by the natural selection process. Perhaps such a breakthrough is possible, but it will require more knowledge and cleverness than fixing a broken wing.

The ability of a group to function adaptively, like flight in birds, is a remarkable and complex adaptation. It is remarkable at the scale of face-to-face groups and even more remarkable at the scale of large modern societies. When we criticize a religion or other social system for failing to perform better or to expand its moral circle still wider, we often implicitly assume that the problem is like a broken wing with an easy solution. If only those Christians were less hypocritical about forgiveness, Christian society would run better and at a larger scale. I suggest that this way of thinking, however well intentioned, is misinformed and ultimately unproductive. Improving the adaptedness of society may require appreciating the adaptive sophistication that already exists. An evolutionary perspective can contribute to this understanding.

I cannot claim too much credit for the evolutionary perspective because Pagels (1995) and her colleagues have already grasped the functional and locally adapted nature of the Gospels without ever breathing the world "evolution." Many evolutionary biologists got it wrong when they rejected group selection and at least one branch of religious scholarship got it right. Nevertheless, at the end of the day we can look forward to a unified theory of human behavior that begins with evolution at one end and offers a satisfying explanation of religion at the other.

This theory might seem like a triumph for science but a defeat for religion in the truest sense of what it means for the religious believer. Perhaps I am a dreamer, but even here I find grounds for optimism. Recall Justin, who saw in Christianity "something beyond nature, a miracle; somehow these people had tapped into a great, unknown source of power." What Justin saw was beyond the human nature that was imagined in Roman times and did tap into a great unknown source of power. It was a new cultural structure that changed the course of human history, arguably for the better. Part of the religious temperament is to imagine a future vastly better than the present. In purely scientific terms this means finding a new cultural structure that is as miraculous in our eyes as Christianity was to Justin. I do not know if such a structure exists, but can anyone prove that cultural evolution has already run its course, that all symphonies have been written and all structures built? I think not.

CHAPTER 7

UNIFYING SYSTEMS

> Cultures are defensive constructions against chaos, designed to reduce the
> impact of randomness on experience. They are adaptive responses, just as
> feathers are for birds and fur is for mammals. Cultures prescribe norms,
> evolve goals, build beliefs that help us tackle the challenges of existence. In
> so doing they must rule out many alternative goals and beliefs, and thereby
> limit possibilities; but this channeling of attention to a limited set of goals
> and means is what allows effortless actions within self-erected boundaries.
> —Csikszentmihalyi 1990, 91

This passage claims for culture in general what I have tried to show for
religion in particular. I like it in part because its author does not study
culture for a living. He is a distinguished psychologist best known for his
research on individual experience. The passage merely reflects what for
him seems common sense.

If the formal study of culture ever converges upon this common-sense
view, as it should, at least three transformations will have taken place. First,
the study of culture will be solidly evolutionary. Second, much of the evolution will be acknowledged to take place at the group level; you can't
talk about cultures having feathers and fur without talking about group
selection. Third, human nature will be seen as something that evolves
rather than as something we are stuck with. After all, nature evolves, so
why not human nature? The first impulse of many evolutionary biologists
would be to say that human nature does evolve, but too slowly to make
a difference. They are thinking of genetic evolution, a limitation that van-
ishes as soon as genetic and cultural evolution are properly integrated with
each other. By analogy, virtually all mammals, from mice to giraffes, have

exactly seven neck vertebrae. I don't know why this trait is so conservative in the mammalian line, but it has not prevented some mammals from evolving very long necks and others from evolving very short necks. Human traits can similarly be constant in some respects, such as the hypothetical innate psychology discussed in chapters 1 and 6, but offer so much flexibility in other respects that cultures can evolve to be as different from each other as mice are from giraffes, with new forms possible in the future that can scarcely be imagined in the present. It is ironic that evolutionary theories of human behavior so often give the impression of an incapacity for change.

In the Introduction I stated that understanding religion requires answers to questions that go beyond religion. Religion is a large subject, but the ideas we have been discussing have an even wider field of application. In this final chapter I will try to gather some of the most important themes and expand my scope from the "tree" of religion to the "forest" of society in general.

A GENERAL THEORY OF UNIFYING SYSTEMS

The word religion is derived from the Latin "religio," which means "to unite or bind together" Related words used outside the context of religion are "religate" (to bind together or unite) and "ligature" (the act of tying or binding up). These meanings reflect the essence of the thesis of this book, like a hidden clue that was not discovered until the very end. However, religions are not the only systems that unite people into adaptive groups. I could have written a book on political organizations, business organizations, military organizations, sports teams, family groups, secular intellectual traditions, or even diffuse cultures as adaptive units. We therefore need to develop a general theory of unifying systems of which religion is a special case. The general theory should distinguish religion from other unifying systems, but it also ought to show what religion shares in common with other unifying systems, including those that appear highly nonreligious in other respects.

FUNCTION AND FUZZINESS

The definition of religion has been debated for as long as the subject has been studied. Stark and Bainbridge (1985, 1987) and others have argued for belief in supernatural agents as the defining criterion, as opposed to Durkheim's distinction between the sacred and profane. Durkheim's definition seems too broad, including but not confining itself to what we think of as religious. Patriotic people regard their flag as sacred but does that make government a religion? On the other hand, Stark and Bainbridge's definition seems too narrow, excluding major religions such as Buddhism, whose founder insisted that he was merely awake and had found a path to enlightenment without the help of any gods. Stark and Bainbridge contend that Buddhism as actually practiced is chock full of gods and therefore is encompassed by their definition. More generally, however, defining religion purely in terms of supernatural agents seems shallow in addition to narrow.

Past debates have tended to ignore two of the most important criteria for defining any class of objects, and therefore religion, which can be labeled function and fuzziness. In chapter 1, I described functionalism and nonfunctionalism as radically different modes of thought that must be used in the appropriate contexts. It is absurd to think of the moon in functional terms and equally absurd to think of a machine or well-adapted organism in nonfunctional terms. The functional status of an object is invariably reflected in its definition. A man-made functional object such as a toothbrush is defined as "a brush for cleaning the teeth." A biologically evolved functional object such as a tooth is defined as a "hard bony structure used for seizing and chewing food and as a weapon." Human groups with functions are also invariably defined in functional terms. A jury is defined as "a body of persons sworn to inquire into and test a matter submitted to them and to give their verdict according to the evidence presented." An army is defined as "a body of men organized for war." As an instructive exercise, I encourage the reader to try to define these objects without referring in any way to their functions. It follows that if religion is a biologically and culturally evolved adaptation, its functional status must be reflected in its definition.

Fuzziness is the degree to which a set of objects (such as the set named "religion") can be distinguished from other sets. The implicit goal of many definitions is to minimize fuzziness. Classical set theory in mathematics embodies this goal by making membership in a set an all-or-none property. Minimizing fuzziness is often useful for practical purposes, as when an anti-aircraft gunner wants to distinguish the planes of its own side from those of the enemy. However, some sets are intrinsically fuzzy. Democracy does not exist as an all-or-none category but rather in degrees. Species are a notoriously fuzzy set of biological objects that defy all-or-none categorization. Understanding intrinsically fuzzy sets requires acknowledging and studying their fuzziness as an interesting property in its own right. A relatively new branch of mathematics and computer science is devoted to the study of fuzzy sets (Kosko 1993; Klir, St. Clair, and Juan 1997), with important implications for the social sciences (Ragin 2000).

Let us return to the definitions of Durkheim and Stark with function and fuzziness in mind:

> A religion is a unified system of beliefs and practices relative to sacred things, that is to say, things set apart and forbidden—beliefs and practices which unite into one single moral community called a Church, all those who adhere to them. (Durkheim [1912] 1995, 44)

> Religion consists of very general explanations that justify and specify the terms of exchange with a god or gods. (Stark 1999, 270)

Durkheim's definition reflects his functional orientation. Religion is a system with a purpose—to unite a human group into a single moral community—in which sacredness features as an essential mechanism. The fuzziness of this definition is a liability if we are trying to minimize fuzziness, but it may also be appropriate if religion is an inherently fuzzy set of objects. Without wishing to vindicate Durkheim in all respects, I think this definition is on the right track because it acknowledges the functional nature of religion.

In contrast, Stark's definition reflects both the byproduct side and the psychological side of his own formal theory of religion. As we have seen, this theory can be faulted on both counts; it ignores functionalism, and its psychological side serves primarily as a metaphor to justify the principle of utility maximization. The accompanying definition must therefore be evaluated on the basis of much more than the presence of gods in Buddhist thought. Even if supernatural agents did turn out to be the defining criterion of religion, we would need to know how the gods function and what replaces them in other unifying systems. Rather than trying to use a definition that cordons off religion from other social systems, we need to understand what all unifying systems have in common and why they vary in ways that impel us to categorize some as "religious," others as "political" and so on, even in the absence of sharp boundaries.[1]

THE COMPARATIVE STUDY OF UNIFYING SYSTEMS

Before we compare human social systems, it is important to remember that unifying systems exist throughout nature. The same theory that explains human groups as adaptive units also explains social insect colonies, individual organisms, and even the origin of life itself as unified groups of interacting molecules that evolved by group-level selection. Chapter 1 began with a passage from the Hutterites, comparing their religious communes to single organisms and bee colonies. As amazing as it may seem, this comparison is fully warranted scientifically. There is much to be gained from the comparative study of unifying systems at this broadest of possible scales.

As we narrow our scope to the comparative study of human social organizations, we need to begin with our own evolutionary history, much of it spent in small social groups approximately the size of modern hunter-gatherer societies. Superficially these groups do not appear favorable for group-level selection. Based on genetic variation alone, we would expect within-group selection to prevail in most cases. However, phenotypic variation within and among human groups is radically different than genetic variation and makes group selection a very strong force indeed. Remember that phenotypic variation is all that natural selection ever "sees." Genetic evolution remains important, but to appreciate its significance we must go beyond the narrowest brand of genetic determinism, so common as a simplifying assumption in evolutionary models, in which genes code directly for behaviors. Instead we must begin to model what we know to be true: genes code for psychological traits that cause people to adopt different behaviors with great (although not unlimited) flexibility and also to alter the environment that determines what counts as adaptive. Perhaps more than any other species, we live in an environment of our own making. Group selection has been a strong force because we have made it so. In more technical terms, human evolution has been a feedback process between traits that alter the parameters of multilevel selection and traits that evolve as a result of the alteration.[2]

When we attempt to study this feedback process in detail, morality emerges as a central phenomenon, as I attempted to show in chapter 1 (see also Sober and Wilson 1998; Boehm 1999). This is an important development because in the past morality and evolution have tended to occupy opposite corners of human thought. Now it appears that they must be studied together, and that even from a purely biological standpoint morality is part of the essence of what it means to be human. One dictionary definition of morality—conformity to the rules of right conduct—nicely summarizes its intimate connection to multilevel selection theory. Conformity eliminates certain kinds of phenotypic variation within groups, regardless of whatever genetic variation may exist. The behaviors that count as right conduct are not genetically determined but depend on open-ended psychological and cultural processes. Once again, some of the psychological processes may be genetically determined, but that is a far cry from assuming that the behaviors are genetically determined. The open-ended nature of right conduct means that phenotypic variation will exist between groups, even as it is reduced within groups by pressures of conformity. By its very nature, morality shifts the balance between levels of selection in favor of group selection (Boehm 1999).

The social sciences are full of scenarios about the lives and minds of our ancestors prior to the advent of civilization. Rousseau imagined a noble savage corrupted by society. Hobbes imagined a brutish savage that must be tamed by society. Freud imagined a guilty savage whose patricidal act somehow became embedded in racial memory. Economists imagine a selfish savage, sometimes even referred to as *Homo economicus*, who becomes civilized only by appealing to his self-interest. It is worth asking why these origin myths are necessary when they have no more basis in fact than the garden of Eden. My guess is that they play a practical role in the belief systems that create and sustain them, much as the distorted versions of history in the Four Gospels. In any case, we are on the verge of being able to replace them with a more authentic picture of the lives and minds of our ancestors. More work is required to refine the concept of guarded egalitarianism and the innate psychology that supports it, but its rough outline provides a starting point for the study of all modern social institutions, religious and otherwise (Bowles and Gintis 1998).

For me, this is the essence of Turner's (1969) distinction between structure and communitas, described in chapter 2. Human societies become structured in many different ways and often become highly stratified. Behind all of them, however, is a strong moral sentiment that society must work for all its members from the highest to the lowest. I interpret this spirit of communitas as the mind of the hunter-gatherer, willing to work for the common good but ever-vigilant against exploitation. In small groups the spirit of communitas results in egalitarian societies with an absence of leaders. In larger groups, which must become differentiated to function adaptively, the spirit of communitas serves as a kind of moral anchor. A large society is robust to the extent that its structure fulfills the spirit of communitas. When it fails, as it so often does from the temptation to benefit oneself at the expense of others within the society, members withdraw their commitment and work to destroy what they previously supported. By a slow and halting process, marked by frequent reversals and by no means destined to move forward, social structures evolve that work roughly for the common good. One important design feature of a robust society is bi-directional control. Leaders and other authorities are required to coordinate action in a large society, but they must also be controlled to prevent abuse of their power. According to Boehm (1999), the anthropologist who has been most influential in developing the concept of guarded egalitarianism, even the most powerful chiefs of traditional societies were ultimately accountable to their so-called subjects.

In this book I have tried to show how a theory of innate psychology can inform the study of religion and how it can clarify, rather than deny, the essential role of culture. The same theory can be applied to any other modern human social organization. As one example, military groups have a strong element of bi-directional control that lies beneath their superficially hierarchical structure. Military leaders gain the respect of their "followers" by making wise decisions and especially by sharing the same risks to which they are exposing others. There is a reason why officers lead their men rather than herding them into battle.[3] A leader who loses the respect of his men stands a good chance of being shot in the back. Richerson and Boyd (1999) explain variation in the fighting ability of American, German, and Russian troops during World War II with the same evolutionary principles that I have used to explain religion.[4] As another example, economists such as Bowles and Gintis (1998) are starting to explore the implications of guarded egalitarianism and its underlying psychology for economic theory and public policy. *Homo economicus* may someday be replaced by a more accurate and scientifically grounded conception of human nature. There is every reason to include religion along with other modern social organizations in this emerging synthesis rather than encapsulating it with definitions that set it apart.

Symbols and Sacredness

For two millennia, Western civilization has imagined people as categorically different from and vastly superior to other animals. The list of supposedly unique human attributes has been almost endless, encompassing language, tool use, intelligence, morals and aesthetics. Ever since Darwin rudely knocked our species off its pedestal, scientists have been trying to discover the real similarities and differences between ourselves and other species. Understandably, much of this research gives the impression that we are not fundamentally different than other animals. The field of evolutionary psychology in its current form, for example, attempts to predict the behavior and minds of people with the same models that work for primates, lions, and ground squirrels.

Given all the false claims about human uniqueness in the past, it may seem surprising that some legitimate claims are emerging, especially from the field of neurobiology. I still remember my surprise when I first read *The Symbolic Species* by Terrence Deacon (1998), which I have mentioned repeatedly throughout this book. He clearly had a commanding grasp of the subject matter, yet he was proposing that the human mind is virtually unique in its capacity for symbolic thought. He even had a clever way of reconciling human uniqueness with the continuity required for any evolutionary argument. According to Deacon, thinking symbolically doesn't require an especially large brain or even a different brain than that possessed by our primate ancestors. In fact, it is actually possible to teach a chimpanzee or bonobo to think symbolically, more like us than their own kind. The problem is that it requires an arduous training process that has no counterpart in nature. Moreover, symbolic thought interferes with more basic forms of associative learning that are adaptive in natural environments. Symbolic thought is like a lofty peak in an adaptive landscape that can be climbed only by first crossing a valley of low fitness. What made humans unique was a natural environmental context that made symbolic thought adaptive in its initial stages, allowing us, and us alone, to cross over to the new adaptive peak.

Deacon's thesis is speculative and may prove wrong in many of its details. I like animals and will be delighted if other species are admitted to the symbolic thought club. What intrigues me most about Deacon's thesis is not the claim about human uniqueness, but the suggestion that symbolic thought might be an essential element of social behavior, at least in humans and perhaps in other species as well. Recall Durkheim's statement that "in all its aspects and at every moment of history, social life is only possible thanks to a vast symbolism" ([1912] 1995, 233). This statement may be ninety years old and well worn in various branches of the social sciences, but it is brand new against the background of modern evolutionary theories of social behavior, including human social behavior. It often seems as if the integration of biology and the social sciences is a one-way street, more a conquest by biology than a fertile interchange. Here is a case where the influence needs to flow the other way.

How can symbols be incorporated into evolutionary theories of social behavior? The first requirement is for them to have an effect on behaviors that in turn influence survival and reproduction (Richerson and Boyd

1985). It is here that the concept of sacred becomes joined to the concept of symbol. Sacred symbols command respect; they dictate behavior. To regard something as sacred is to subordinate oneself to it, to obey its demands. In contrast, to regard something as profane is to subordinate it to oneself, to use it for one's own purposes. Sacred symbols organize the behavior of the people who regard them as sacred. This may sound (and be) mundane from a social scientific perspective, but it provides a workable starting point for incorporating the concepts of symbols and sacredness into evolutionary theory.

It may help to recall the sacred office of the leopard-skin chief among the Nuer to make our discussion less abstract. Imagine a man who has just slain another racing to the compound of the leopard-skin chief with the kinsmen of the deceased thundering in pursuit. Avenging the death of a kinsman is one of the strongest biological and cultural (for the Nuer) imperatives that can be experienced. Nevertheless, they stop at the boundary of the compound, not because of physical might but because of a social convention conveyed through the use of a symbol. Of course, it is likely that the symbol is backed by physical might. If the avenging kinsmen had invaded the compound, they might have been duly punished by other members of the tribe for their sacrilege. However, this likelihood merely makes the concept of sacred symbols fit comfortably within the framework of guarded egalitarianism. We already appreciate that human groups easily form into moral communities that enforce their sense of right conduct. We merely need to add that sacred symbols provide a mechanism for representing a moral system and putting it into action.

Surely we are near the heart of religion when we talk about sacred symbols. But have we left the heart of other social organizations? I think not. If symbolic thought is as central to human evolution and mentality as Deacon thinks, it must lie at the center of all human social life. All of our behavior all of the time is organized by a vast symbolism, much as Durkheim claimed.

FACTUAL AND PRACTICAL REALISM

One hallmark of religion is its otherworldly nature that to a nonbeliever seems detached from reality. In addition to their gods, religions make the real world otherworldly by altering the nature of the people that actually did exist and the events that took place among them. As we have seen, the New Testament Gospels make poor history, not simply because memory fades with time, but because historical veracity was subordinated to the symbolic use of narratives about people and events to motivate action.

Those who regard themselves as nonreligious often scorn the otherworldliness of religion as a form of mental weakness. How could anyone be so stupid as to believe in all that hocus-pocus in the face of such contrary evidence? This stance can itself be criticized for misconstruing and cheapening a set of issues that deserves our most serious attention as scientists and intellectuals.

In the first place, much religious belief is not detached from reality if the central thesis of this book is correct. Rather, it is intimately connected to reality by motivating behaviors that are adaptive in the real world—an awesome achievement when we appreciate the complexity that is required to become connected in this practical sense. It is true that many religious beliefs are false as literal descriptions of the real world, but this merely forces us to recognize two forms of realism; a factual realism based on literal correspondence and a practical realism based on behavioral adaptedness. An atheist historian who understood the real life of Jesus but whose own life was a mess as a result of his beliefs would be factually attached to and practically detached from reality.

In the second place, much religious belief does not represent a form of mental weakness but rather the healthy functioning of the biologically and culturally well-adapted human mind. Rationality is not the gold standard against which all other forms of thought are to be judged. Adaptation is the gold standard against which rationality must be judged, along with all other forms of thought. Evolutionary biologists should be especially quick to grasp this point because they appreciate that the well-adapted mind is ultimately an organ of survival and reproduction. If there is a trade-off between the two forms of realism, such that our beliefs can become more adaptive only by becoming factually less true, then factual realism will be the loser every time (Wilson 1990). To paraphrase evolutionary psychologists, factual realists detached from practical reality were not among our ancestors. It is the person who elevates factual truth above practical truth who must be accused of mental weakness from an evolutionary perspective.

In the third place, disparaging the otherworldly nature of religion presumes that nonreligious belief systems are more factually realistic. It is true that nonreligious belief systems manage without the gods, but they might still distort the facts of the real world as thoroughly as the Four Gospels of the New Testament. We know this is the case for patriotic versions of history, which are as silly and weak-minded for people of other nations as a given religion for people of other faiths. Many intellectual traditions and scientific theories of past decades have a similar silly and purpose-driven quality, once their cloak of factual plausibility has been yanked away by the hand of time. If believing something for its desired consequences is a crime, then let those who are without guilt cast the first stone.

The study of evolution is largely the study of trade-offs. Becoming better in some respects requires becoming worse in others, which in part explains why life consists of a diversity of forms rather than one all-purpose species. The proper and intellectually respectful way to approach factual and practical realism is as a trade-off. Factual knowledge of one's physical and social environment is useful for many purposes. All people (including the most "primitive") have this kind of knowledge and can express it in at least some contexts, such as the Balinese farmer discussing pest control quoted in chapter 4. However, it appears that factual knowledge is not always sufficient by itself to motivate adaptive behavior. At times a symbolic belief system that departs from factual reality fares better. In addition, the effectiveness of some symbolic systems evidently requires believing that they are factually correct. Constructing a symbolic system designed to motivate action is a substantially different cognitive task than gaining accurate factual knowledge of one's physical and social environment. Somehow the human mind must do both, despite the fact that they partially interfere with each other.

How this trade-off is managed is anyone's guess, because to my knowledge the subject has seldom been formally conceptualized in this way (Wilson 1990, 1995). One possibility is that both forms of knowledge evolve by blind evolution rather than by cognitive processes. People will believe anything, and those with the right mix of factual and practical realism replaced those who erred on either side. Another possibility is that every individual has highly sophisticated mental modules for acquiring factual knowledge and for building symbolic belief systems, which are subordinated to each other depending upon the context. Perhaps there is a hard-headed factual realist playing a quiet role within every religious zealot. A third possibility is a division of labor among individuals. Stark and Bainbridge (1985) observe that atheists have existed throughout history, including in so-called "primitive" societies. Perhaps what seems to be an adversarial relationship between believers and nonbelievers in fact represents a healthy balance between factual and practical realism that keeps social groups as a whole on an even keel. Instead of thinking of religion as uniquely otherworldly and weak-minded, we need to discover how the trade-off between factual and practical realism is managed in all human social organizations and how it can be managed better in the future.

Science and Society

So far we have made progress sketching a general theory of human unifying systems but less progress identifying anything unique about religion. Perhaps we will have more luck if we try to find something unique about science. It is interesting to speculate that science is unique in only one respect: its explicit commitment to factual realism. Virtually every other human unifying system includes factual realism as an important and even essential element but subordinates it to practical realism when necessary. Only science, in the most idealistic sense of the word, attempts to eliminate the trade-off by taking the advancement of factual knowledge as its explicit purpose. One could say that factual knowledge is the god of science. In every other respect science might be just like all other unifying human social organizations, including religion.

A number of implications follow from this speculation. First, science emerges as an unnatural act. The human mind is probably far better at subordinating factual to practical realism than the reverse. The ideal of the true scientist, who weighs only the facts without regard to the practical consequences, is about as attainable as the ideal of Jesus, Muhammad, or Buddha. The best that science can do as a social organization is to implement a system of beliefs and practices that steers people toward the ideal. Of course, we know that scientific culture is packed with sacred symbols, self-glorifying statements, and reasonably effective social control mechanisms. It might seem that I am disparaging science by comparing it to religion in this way. On the contrary, I think that science might profit by becoming more religious along certain dimensions, as long as it remains nonreligious with respect to its stated goal of increasing factual knowledge. Science needs an effective structure that implements a spirit of communitas as much as any other human unifying system.

It is likely that practical realism must be anchored in factual realism to remain practical over the long term. It therefore makes sense for a large differentiated society to include an organ—to speak in unabashedly functionalist terms—for which the acquisition of factual knowledge is the highest value. However, it is the nature of the trade-off that factual realism by itself does not constitute practical realism. It follows that the values of scientific society do not suffice for the society as a whole. They must be supplemented with other values that place a greater emphasis on practical realism and that hopefully apply to all members of the society as moral equals. This is important because the glorification of science often makes it appear as an attractive model for all aspects of society to emulate, to the consternation of those who think that something precious would be lost were science to become the dominant and exclusive model. Thinking of science in relation to society in explicitly functional terms helps us to avoid this elementary error.

BEAUTY AND UTILITY

Religions and other symbolic systems designed to motivate action have a beauty and power that seem to wilt when explained in functional terms. Surely, someone who thinks he is playing a decisive role in a cosmic battle between good and evil would be disappointed to learn that he is merely playing a role in the economy of his church. Is it possible to openly acknowledge the functional nature of a unifying system and retain a sense of motivating beauty?

The attitude that usefulness is ugly is actually quite peculiar. Evolution offers a theory of aesthetics that predicts just the opposite. Far from a uniquely human capacity, a sense of beauty is probably the emotional output of ancient brain mechanisms designed to evaluate the fitness associated with features of the environment. These mechanisms are computationally sophisticated but operate beneath conscious awareness. Just as the act of seeing appears effortless but in fact belies an enormously sophisticated cognitive process, we are emotionally attracted to features of the physical and social environment that are likely to increase our fitness, which we experience as beautiful. This theory has enjoyed considerable success predicting the features of sexual partners and natural landscapes that are regarded as beautiful (reviewed by Thornhill 1998). In the present context, we can predict that social groups will be regarded as very beautiful indeed when they offer the ingredients of survival and reproduction that their members would otherwise lack.

Some of the most beautiful and moving elements of religion come not from cosmic struggles and invisible gods but from the vision of a better life on earth. Recall the famous passage from Matthew (25:35–40) quoted in chapter 4, which gives thanks for the simplest gifts imaginable during a time when a compassionate society was a miracle. To illustrate the beauty and motivating power of function, and in the spirit of thinking of science as different from religion in only one respect, I offer the following humble parable of my own.

Suppose that you are an airplane enthusiast who has built your own airplane. Building it was as much fun as flying it and you spend hours in admiration, not only because it is your handiwork but also because of its gorgeous design: its sleek aerodynamic shape and gleaming engine that miraculously cause such a heavy object to rise into the heavens. As you run your hand over the engine, so complex yet so comprehensible for those who know it, something catches your eye. What is it?

Let's say that it is a socket wrench, thoughtlessly left by the mechanic who serviced the plane yesterday in preparation for your next flight. Thoughtless isn't the right word—fatal is more appropriate, since the wrench might well have caused the engine to fail in mid-flight. Horrified, you remove it and vow to teach the mechanic a lesson he will never forget. Or . . .

Let's say that it is an enormous gob of gum thoughtlessly left by the mechanic. This gob isn't life threatening but it is disgusting, not just because it was in his mouth but because of where he stuck it, like someone spitting in a cathedral. Furious at what others might regard as a trivial act, you vow to change your mechanic. Or . . .

Let's say that it is a picture, drawn by the mechanic with a marker on one of the engine parts. He must have done it during his lunch break. Furthermore, he must have gone to art school before becoming a mechanic because the picture is quite impressive. Perhaps you would still angrily erase the picture because it is out of place and offends your aesthetic appreciation of the engine, which doesn't need a picture to make it beautiful. Perhaps you like it enough to leave it. Perhaps you even like it so much that you invite the mechanic to paint your whole airplane. Regardless, for you the beauty of the picture can't hope to rival beauty of the airplane. This fact strikes you with special force during your next flight when you encounter a thunderstorm. A downdraft sends you spiraling toward the earth and rattles the wings to their very roots, but the structure holds as you nimbly pull out of your tailspin. Even a paint job by Picasso would pale against that kind of beauty. Or . . .

Let's say that the mechanic has fiendishly disconnected your entire engine and reworked it into a piece of modern art. Attached to the manifold is a neatly typed label that reads "This engine represents the decline of Western Civilization." Now your thoughts turn homicidal. You contemplate kidnapping your mechanic, locking him into the cockpit of your airplane, towing it into the skies with another airplane that works and releasing it at an altitude of 30,000 feet. Taped to the steering wheel of your plane is the mechanic's own note with "Western Civilization" crossed out and "my plane with you in it" written in.

Thus begins and ends my career as an inspirational writer. It is sufficiently motivating for me to think of society as an aircraft of our own making, which can fly effortlessly toward the heavens or crash and burn, depending upon how it is constructed. I am also encouraged by some of the examples of religious belief that we have encountered in the pages of this book, which combine a hard-headed factual realism with the profound respect for symbols embodied in the word "sacred." Like the Nuer tribesman and Balinese farmer, let us know exactly what our unifying systems are for, and then pay them homage with overflowing belief.

NOTES

INTRODUCTION

1. I thank philosopher Eric Dietrich for the elegant wording of these definitions.

CHAPTER 1

1. Other knowledgeable accounts of multilevel selection theory are provided by Boehm (1999), Frank (1998), Maynard Smith and Szathmary (1995), Michod (1999a), Seeley (1995), and special issues of two journals; *The American Naturalist* 150 (1997, supplementary issue) and *Human Nature* 19, no. 3 (1999).

2. The words "altruistic" and "selfish" have many meanings, both in everyday language and as scientific terms. Evolutionary biologists define altruism entirely in terms of fitness effects, without reference to how individuals think or feel. In addition, multilevel selection theory defines altruism in terms of relative not absolute fitness. A behavior is selfish when it increases the fitness of the actor, relative to other members of its group. A behavior is altruistic when it increases the fitness of the group, relative to other groups, and decreases the relative fitness of the actor within the group. See Sober and Wilson 1998 and references cited therein for a more complete discussion of altruism, including its psychological meanings.

3. The example shows that the calling behavior increases in frequency in the global population over the short term. However, more information is required to know whether the calling behavior will evolve over the long term. If the groups remain permanently isolated from each other, the local advantage of selfishness will run its course within each group and drive altruism extinct. There must be a sense in which the groups compete with each other in the formation of new groups, although the competition need not be direct. Also, the formation of new groups must include a mechanism for creating the extreme variation among groups that is part of the example. These elements are broadly referred to as "population structure" and are discussed in detail in Sober and Wilson 1998.

4. Average fitness across groups can be useful in a variety of contexts and is fallacious mostly when used to argue against group selection. However, since the major theoretical frameworks that average fitness across groups were developed specifically as an alternative to group selection (e.g., inclusive fitness theory, evolutionary game theory, and selfish gene theory), the averaging fallacy is a major source of confusion in evolutionary thinking.

5. Some social groups have discrete boundaries while others do not. Sharp boundaries exist for people in apartment buildings and parasites in hosts but not for people in neighborhoods or plants in fields. In all of these cases, however, social interactions are localized and influence the fitness of a small subset of the population. It is the localized nature of social interactions and not sharp boundaries that form the basis of multilevel selection theory.

6. Readers familiar with evolutionary game theory should try this exercise on cooperative strategies such as "tit-for-tat (T)" interacting with selfish strategies such as "all-defect (D)" in which the groups are the pairs of interacting individuals ($N = 2$). D gets a higher payoff than T within mixed groups, while TT groups get a higher combined payoff than TD or DD groups. It is interesting that Anatol Rapoport, who submitted the T strategy to Robert Axelrod's famous computer simulation tournaments (Axelrod 1980a, b) appreciated its locally altruistic nature and criticized his own colleagues for thinking of it as an individually selfish strategy. As he put it, "The effects of ideological commitments on interpretations of evolutionary theories were never more conspicuous" (Rapoport 1991, 92).

7. Achieving the middle ground is proving to be a slow process. Many articles, textbooks, and popular books on evolution still portray the age of individualism as alive and well. Major figures who built their reputations on the rejection of group selection have been slow to change their minds. The adage "science progresses . . . funeral by funeral" is perhaps too pessimistic, but readers will need to rely on their own judgment more than usual until the "authorities" reach a consensus.

8. Some authors prefer to make the concept of organism more restrictive than my somewhat casual usage (Sterelny and Griffiths 1999). Thus, a group that behaves adaptively in a single context, such as predator defense, may not qualify as an organism because it does not behave adaptively in other contexts. The question of whether and how many adaptations become bundled together in a single group is certainly interesting and important, regardless of how we use the word organism. Also, both the casual and restrictive definitions rely centrally on multilevel selection theory to explain the concept of organism.

9. Dawkins's concept of "selfish genes" involves averaging the fitness of genes across individual genotypes, just as I averaged the fitness of individuals across groups in my bird flock example. It is therefore not an argument against either individual selection or group selection, but merely newspeak for "anything that evolves by natural selection at any level." For example, a gene for altruistic calling behavior that is selectively disadvantageous within groups and evolves by group selection is nevertheless "selfish" merely because it replaces the noncalling gene

in the total population. Dawkins developed the selfish gene concept as an argument against group selection, and it was regarded as such for years before its irrelevance became recognized and eventually acknowledged by Dawkins himself in *The Extended Phenotype* (Dawkins 1982).

10. Just as outlaws are never entirely eliminated within human social groups, individual organisms are turning out to be less harmonious than originally thought, with a variety of "outlaw" elements that replicate themselves at the expense of their collectives. This fascinating subject is known as "intragenomic conflict" in the evolutionary literature (reviewed by Pomiankowski 1999).

11. The averaging fallacy is especially common in evolutionary accounts of human behavior. The usual pattern is for group selection to be rejected after a brief discussion, after which cooperative aspects of human society are acknowledged but described in individualistic terms. The existence of groups and intergroup competition are often emphasized as important factors in human evolution without raising the issue of group selection in the author's mind. I have analyzed three prominent examples of the averaging fallacy in detail: Blurton-Jones's (1984, 1987) model of tolerated theft as an explanation of food sharing in hunter-gatherer societies (Wilson 1998a), Richard Alexander's views on human evolution (Wilson 1999a), and Robert Wright's (2000) recent book *Nonzero*, which portrays human society as organismic on a vast scale without breathing the word group selection (Wilson 2000).

12. The vast majority of evolutionary models of social behavior, including kin selection models, assume a one-to-one relationship between genes and behaviors. Strict genetic determinism is assumed, not because it is realistic but because it simplifies the mathematics. The hope is that more complex and realistic connections between genes and behavior will not alter the basic results of the simpler models. It turns out that this hope is unfounded, creating new possibilities for multilevel selection that are only beginning to be explored (Wilson and Kniffin 1999).

13. Both examples of hunter-gatherer moral systems described in this chapter are from a survey of twenty-five cultures chosen at random from the Human Relations Area File, an anthropological database designed for cross-cultural research and discussed in chapter 5 of Sober and Wilson 1998. Random sampling insures that the cultures in the sample are representative of all the cultures from which the sample was drawn. In chapter 4 I will employ the same technique to obtain a random sample of world religions.

14. According to Howell's (1984) ethnography, the Chewong are thoroughly nonviolent even in their punishment of transgressions. Usually, the transgressor simply takes the hint and moves away, but it is clear that social disapproval is the cause of the voluntary exile.

15. See Gaulin and McBurney 2001 for an excellent general discussion of domain-specific learning mechanisms.

16. My discussion of cultural evolution must necessarily be kept brief and focused on the central themes of this book. Influential treatments of cultural evolu-

tion include Boyd and Richerson 1985, Cavalli-Sforza and Feldman 1981, Dunbar et. al. 1999, Durham 1991, Lumsden and Wilson 1981, Sperber 1996, and Tomasello 1999.

17. The term "evolutionary psychology" should refer broadly to the study of psychology from an evolutionary perspective. Unfortunately, the term has become associated with a rather narrow school of thought that is controversial even among evolutionary biologists interested in psychology. Books on self-described evolutionary psychology include Barkow, Cosmides, and Tooby 1992, Cartwright 2000, Cosmides and Tooby 2001, Buss 1999, and Gaulin and McBurney 2001. Books by qualified evolutionary biologists who are critical of self-described evolutionary psychology include Chisholm 1999, Heyes and Huber 2000, Scher and Rauscher forthcoming, and Sterelny and Griffiths 1999. I think it is important to recognize the virtues of self-described evolutionary psychology while expanding its horizons (Wilson 1994, 1999b, 2002).

18. Tooby and Cosmides (1992) acknowledge a process of transmitted culture in addition to what they call "evoked culture," but transmitted culture plays almost no role in their conceptual framework.

19. It is notable that G. M. Edelman, who received a Nobel Prize in 1972 for his work on the immune system, went on to develop a similar conception of the brain, which he termed "neural darwinism" and even "neuronal group selection" (Edelman 1988; Edelman and Tonomi 2001). This approach to human mentality has not yet been integrated with self-described evolutionary psychology.

20. It is easy to imagine cultural transmission processes that favor within-group selection rather than between-group selection. For example, the transmission rule "copy the most successful person in your group" would cause everyone within a group to emulate selfishness. The fact that cultural evolution in our species facilitates group selection probably reflects the fact that cultural evolution is itself a product of genetic evolution. This example also illustrates why familiar sounding terms such as "imitation" need to be clarified before they can be predictive. Imitation at a small (within groups) scale causes selfishness to be emulated while imitation at a large scale (across groups) can cause altruism and other group-advantageous behaviors to be emulated. There are many kinds of imitation, only some of which are observed in our species. See Wilson and Kniffin (1999) for an explicit model of the genetic evolution of cultural transmission rules.

21. Adaptive landscapes are a compelling visual metaphor but have proven difficult to formalize. Sewall Wright, who popularized the metaphor, employed at least three different versions that are incompatible with each other (Provine 1986).

22. Sewall Wright thought that group selection could turn evolution in rugged adaptive landscapes into a creative process. Although a single population might become "trapped" on a given peak, multiple populations might occupy different peaks and those occupying the highest peak might eventually out-compete the others. The ruggedness of adaptive landscapes and the feasibility of Wright's

"shifting balance theory" are still hotly debated (Coyne, Barton, and Turelli 1997, 2000; Wade and Goodnight 1998). Boyd and Richerson (1992) have developed a similar idea for social evolution, which they call group selection among multiple evolutionary stable strategies.

23. Some recent books on evolution and religion include Boyer 2001, Burkert 1996, Hinde 1999, Goodenough 1998, Gould 1999, Miller 1999, and Ruse 2000. None of these authors approach religion from a multilevel perspective with an emphasis on group selection, although some discuss cooperation from an individualistic perspective, which I have termed the averaging fallacy. Boyer (2001) treats religion primarily as a byproduct of psychological adaptations that are beneficial in nonreligious contexts. Goodenough (1998), Gould (1999), Miller (1999), and Ruse (2000) discuss how science and evolution can be reconciled with religious belief rather than trying to explain religion from an evolutionary perspective.

CHAPTER 2

1. There is no firm distinction between a theory and a hypothesis. A theory is relatively general and a hypothesis is relatively specific, but both are explanations of a given phenomenon. I refer to my explanation of religion as a hypotheses because it is one of several offered by evolutionary theory, as shown in table 1.1. Stark refers to his rational choice explanation of religion as a theory, and I have decided to preserve his usage in this book. As I will show, Stark's theory needs to be broken into a number of hypotheses similar to those for evolutionary theory.

2. Ironically, adaptationists such as Dawkins (1982) and Williams (1996) tend to regard themselves as reductionistic while critics of adaptationism such as Gould and Lewontin (1979) tend to regard themselves as holistic. Nevertheless, it is the adaptationists who explain organisms without reference to their actual genetic, chemical, or physical make-up. Wilson (1988) distinguishes three different meanings of holism and reductionism in ecology and evolutionary biology that need to be distinguished from each other.

3. Complex systems theory and multilevel selection theory have yet to be fully integrated with each other. It is unlikely that complexity in the absence of selection can explain the existence of functional organization at any level of the biological hierarchy, despite claims to the contrary (reviewed by Depew and Weber 1995). However, complexity can have profound effects on variation and heritability, the two prerequisites for natural selection. When complexity is viewed as a source of heritable variation, it creates new possibilities for multilevel selection that have been previously regarded as highly unlikely on the basis of simple interactions. These ideas are highly relevant to human genetic and cultural evolution but can only be touched upon here. See Wilson (1992), Swenson, Arendt, and Wilson (2000), and Swenson, Wilson, and Elias (2000) for fuller discussions of the issues in the context of selection at the level of biological communities and ecosystems.

4. According to the invisible hand metaphor in economics, human society works well even though its members attend only to their narrow self-interest. One possibility is that individual self-interest invariably leads to well-functioning societies. If the group selection controversy has taught us anything, it should be to distrust this possibility. Alternatively, group selection may have evolved mechanisms that cause individuals to participate in an adaptive economy with no more awareness of their roles than bees performing their waggle dances. The Nobel prize-winning economist F. Hayek (1988) speculated along these lines, properly invoking group selection.

5. Sober and Wilson (1998, chap. 5) discuss the Nuer conquest from a multi-level evolutionary perspective.

CHAPTER 3

1. In chapter 1, I distinguished between religion as it is idealized and religion as it is actually practiced. Idealized religion is close to what one expects on the basis of pure group selection, or so I claim. Religion as practiced often deviates from the ideal, but these deviations tend to be regarded as "corruptions" rather than as a part of the religion itself. It is useful to retain this distinction and to analyze religion in both its idealized and practiced forms. Calvin's catechism is an idealized religion, and I am not so naive as to think that all the citizens in Geneva obediently obeyed its principles. However, it would be equally naive to think that Calvinism had no effect whatsoever on the behavior of the Genevans, as the rest of this chapter will show. The idealized form was designed to establish an organismic community, and the realized form partially succeeded.

2. Populations not only adapt to their environment but also change their environment, resulting in a more co-evolutionary process, as emphasized by Laland, Odling-Smee, and Feldman 2000.

3. Several reviewers advised me to change this example because, according to recent research, spinach is not healthy for its iron content after all! However, this only underscores the problem of trying to justify behavioral prescriptions purely on the basis of known facts.

4. The language of stratified societies (God as a powerful and just Lord) is used just as often as kinship terms (God as father), even though stratified societies are culturally recent in evolutionary terms.

5. I do not mean to imply that religious believers are mindless drones who unquestioningly do as they are told. Individuals also use their religious beliefs as a framework for making personal decisions. Also, in many religious groups the leaders are so highly accountable to the congregation that the decision-making unit comprises the entire group.

6. Although the design features of a religious system such as Calvinism can be evaluated on their own terms (argument from design), it is better to evaluate them on a comparative basis (phenotype-environment correlation). A more comprehensive analysis would include the Christian denominations that were contemporary with Calvinism, in addition to longitudinal studies of how Calvinism

spread relative to other forms of Christianity and how it in turn gave rise to new forms. I hope that this book will encourage more ambitious efforts. See Swanson (1967) for a comparative study of Reformation movements from a sociological perspective.

7. Religious believers might claim that community is one purpose of religion but not the only or even the main purpose. I discuss this issue in terms of proximate and ultimate explanation in chapter 5.

8. The events that caused many religious systems to arise and caused some to prevail over others during the Reformation period can be understood in terms of the evolution of complex systems discussed in note 3 of chapter 2. Complex interactions cause even large social units such as cities and nations to become different from each other. Some variants are unstable and dysfunctional, quickly falling apart, while others are more robust and spread to new locations, either by conquest or imitation. The factors that cause a given social unit to embark upon a given trajectory may be so complex that they will never be fully understood, giving history an unpredictable element. However, it is highly predictable that social units will vary and that variation will have evolutionary consequences. The evolution of complex systems provides an alternative to thinking of cultural evolution in terms of atomistic gene-like units.

CHAPTER 4

1. I mentioned in chapter 2 that the social sciences are an archipelago of disciplines that only partially communicate with each other. Functionalism has been largely, but not entirely, rejected across disciplines. One purpose of this book is to promote conceptual integration within the social sciences in addition to explaining human groups from an evolutionary perspective.

2. Irrigation systems are built and maintained by many traditional societies and provide a good context for studying group-level adaptation (Ostrom 1991). The social organizations that regulate water use are not always religious in nature. I discuss the relationships between religion and other social organizations in chapter 7.

3. Here is an example of a highly specific injunction to contribute taxes: "Reminder that the sacred ruler at Batur [Dewi Danu, the Goddess of the Lake] possesses a congregation of 45 villages. The village of Batur is reminded of what is owed from the possessions held by these villages. They are also reminded of the specific taxes owed to the sacred ruler. They risk the curse of the Deity if they neglect these obligations. And the rice terraces belonging to the sacred ruler, [are] held in trust by the village of Batur. And if they [Batur] do not offer up the rice taxes to the sacred ruler for each yearly temple festival and offer the contributions to the Goddess at Tampurhyang, the people who hold the rice terraces belonging to the sacred ruler will be cursed by the Goddess. May this curse never occur. If the people of the Goddess of Batur do not follow the instructions of the Goddess, and provide the contributions specified, their crops will fail and they will be cursed by the sacred ruler at Batur. And the people of central Bali, if they forget

the holy places at Batur, they will instantly suffer disasters, their works will fail, all that they plant will die, because the Goddess is entitled to the essence of the work of the people of central Bali. This essence is named sasalaranta. Because the Goddess makes the waters flow, those who do not obey her rules may not possess her rice terraces" (from Rajapurana Ulun Danu Batur, quoted in Lansing 1991, 104).

4. Lansing (1991) describes in detail how the Dutch misinterpreted Balinese society by assuming that it was similar to Dutch society. He also describes the exploitative nature of the Dutch occupation of Bali, providing a good example of between-group interactions that differ sharply from within-group interactions.

5. Lansing's (2000) own interpretation of the water temple system emphasizes the concept of self-organization, in which complex systems arise out of elements that follow simple decision rules. As I stated chap. 2., note 3 and chap. 3, note 5, complex systems can have important effects on variation and heritability but are unlikely to explain functional organization in the absence of selection. The process of selection that gave rise to the water temple system of Bali is currently less well understood than the functionality of the system itself.

6. Demographic change in a population depends upon births, deaths, immigration, and emigration. The balance of these inputs and outputs must be positive for any religion to persist, but their relative importance can vary widely. As one extreme example, the Shakers were successful for a brief period of time based purely on immigration and without any births. Based on immigration alone, Judaism is at a large disadvantage compared to proselytizing Christian and Islamic religions, which accounts in part for its minority status. Despite its disadvantage with respect to immigration, however, Judaism has persisted on the strength of the other factors that contribute to demographic growth (high birth rate, low death rate, and low emigration).

7. This provides a vivid example of how cultural variation can differ from genetic variation and can be maintained despite extensive migration between groups.

8. Evolution has frequently been used to justify racist and anti-Semitic doctrines. However, the solution to this problem is not to reject evolution but to vigorously combat its use for inappropriate and unjustified purposes. A similar situation exists for the Christian religion, which has also been used to justify racist and anti-Semitic doctrines. Our own forefathers relied on claims regarding the superiority of the Christian civilization, not the principle of evolution, to justify their policy toward Native Americans. Eliminating racism and anti-Semitism requires a fundamental understanding of why people so easily exclude other people from their moral circles. Evolution in general and multilevel selection theory in particular contribute to this understanding.

9. Although Hamilton (1964) originally developed kin selection theory as an alternative to group selection, he was among the first to change his mind, based on the work of George Price (Hamilton 1975). See Sober and Wilson 1998,

chap. 2, Hamilton 1996, and Schwartz 2000 for accounts of this fascinating period in the history of evolutionary biology.

10. Jews are only one of several ethnic groups that have existed in diaspora situations throughout history. The method of phenotype-environment correlation discussed in chapter 2 predicts that these ethnic groups should converge in many of their adaptive properties, which appears to be the case (Landa 2000).

11. Female infanticide can be adaptive at the group level when success in warfare depends on the number of males. Groups that commit female infanticide do not suffer from a shortage of females because they capture females from other groups. However, when all groups practice female infanticide there is a global shortage of females. This is an example of a behavior that is adaptive at the group level but maladaptive at the metagroup level, just as selfishness is adaptive at the individual level but maladaptive at the group level.

12. Wilson and Sober (2001) develop an argument that shows how social control can promote altruism rather than replacing altruism. This positive relationship is certainly suggested by the high degree of self-sacrifice promoted by religions.

13. As I mentioned at the end of chapter 1, I am trying to evaluate the evidence for religious groups as adaptive units without claiming that religious groups are invariably adaptive. Indeed, the process of natural selection involves many failures for each success, so the existence of failures does not count as evidence against the theory. A survey such as this needs to be interpreted in terms of multilevel theory in general, not group selection alone. I claim that group selection has been a very important force, but not the only force, in the cultural evolution of religions and in the genetic evolution of the psychological mechanisms relevant to religious beliefs and practices.

14. I excluded ethnic religions because they are often so intertwined with the rest of the culture that it is difficult to know where one begins and the other ends. Religions with known starting dates are more easily recognized and studied as relatively discrete cultural units.

CHAPTER 5

1. Social support systems exist for many immigrant groups and not all of them are religious. I discuss the relationship between religious and nonreligious social organizations in chapter 7.

2. The adaptationist program would first attempt to explain the function of these offices in terms of their material benefits to the group and/or individual. If secular benefits cannot be demonstrated, then the offices would be interpreted as a nonadaptive byproduct of a psychological drive for respect that is adaptive only in a larger context. This point illustrates that byproduct theories might explain some aspects of religion, even though they are unlikely to provide a general theory of religion.

3. This distinction is similar to Dennett's (1981) distinction between the intentional stance and the design stance. If you are playing a game of chess with a chess-playing computer, your first assumption should be that it is well designed to play chess. However, if you know exactly how it is designed to play chess, you may be able to exploit some of its foibles.

CHAPTER 6

1. Forgiveness is being defined at the behavioral level, regardless of the psychological mechanisms that cause forgiveness in a proximate sense. The relationship between forgiveness at the behavioral and psychological levels is likely to be complex and requires an analysis comparable to Sober and Wilson's (1998) treatment of altruism.

2. We should also expect important individual differences in these capacities. The nature of individual differences in humans and other animals is discussed by Wilson (1994, 1998c), and Wilson, Near, and Miller (1996).

CHAPTER 7

1. Our tendency to think of religion as a distinct kind of social organization may be influenced by the separation of church and state during very recent times. Needless to say, this is a poor foundation for thinking about religion at a fundamental level.

2. Explicit models of gene-culture coevolution attempt to model this process. See Boyd and Richerson 1985; Durham 1991; Lumsden and Wilson 1981; Wilson and Kniffin 1999. See Laland, Odling-Smee, and Feldman 2000 for a general review.

3. My father has an amusing story from World War II that illustrates this point. He was captain of a small supply ship in the South Pacific and one day his men rigged a tarp to catch rain water. Thinking that the tarp had filled inadvertently, my father proceeded to take a bath in it. Despite the fact that he was captain, in his own words "I thought they were going to kill me."

4. Richerson and Boyd (1999) use the term "crude superorganism" to emphasize both the presence and absence of group-level adaptations. I agree with this characterization but think that the apparent "crudeness" of an adaptation depends on one's background expectations. The functionality of a religious system such as Calvinism or the water temples of Bali is very sophisticated against the background of modern intellectual thought, which is dominated by individualism. In addition, the adaptationist program involves predicting what highly adaptive groups would look like, knowing that real groups will fall short to a greater or lesser degree. For all of these reasons, we should allow ourselves to freely imagine social groups as highly sophisticated adaptive units.

BIBLIOGRAPHY

Abrams, D., and M. A. Hogg. 1990. *Social identity theory: Constructive and critical advances.* New York: Springer-Verlag.

———, eds. 1999. *Social identity and social cognition: An introduction.* Malden, Mass.: Blackwell.

Alcoholics anonymous. 1976. 3d ed. New York: Alcoholics Anonymous World Services.

Alexander, R. D. 1979. *Darwinism and human affairs.* Seattle: University of Washington Press.

———. 1987. *The biology of moral systems.* New York: Aldine de Gruyter.

Allen, N. J., W. S. F. Pickering, and W. W. Miller, eds. 1998. *On Durkheim's "Elementary forms of religious life."* London: Routledge.

Almann, J. M. 1999. *Evolving brains.* New York: Scientific American Library.

Axelrod, R. 1980a. Effective choices in the prisoner's dilemma. *Journal of Conflict Resolution* 24:3–25.

———. 1980b. More effective choices in the Prisoner's Dilemma. *Journal of Conflict Resolution* 24:379–403.

———. 1984. *The evolution of cooperation.* New York: Basic Books.

Axelrod, R., and W. D. Hamilton. 1981. The evolution of cooperation. *Science* 211:1390–96.

Barkow, J. H., L. Cosmides, and J. Tooby, eds. 1992. *The adapted mind: Evolutionary psychology and the generation of culture.* Oxford: Oxford University Press.

Blackmore, S. 1999. *The meme machine.* Oxford: Oxford University Press.

Blurton Jones, N. 1984. A selfish origin for human food sharing: Tolerated theft. *Ethology and Sociobiology* 5:1–3.

———. 1987. Tolerated theft: Suggestions about the ecology and evolution of sharing, hoarding and scrounging. *Social Science Information* 26:31–54.

Boak, A. E. R. 1955. *Manpower shortage and the fall of the Roman Empire in the West*. Ann Arbor: University of Michigan Press.

Boehm, C. 1978. Rational preselection from Hamadryas to Homo Sapiens: The place of decisions in adaptive process. *American Anthropologist* 80:265–96.

———. 1983. *Montenegrin social organization and values: Political ethnography of a refuge area tribal adaptation*. New York: AMS press.

———. 1984. *Blood revenge*. Philadelphia: University of Pennsylvania Press.

———. 1993. Egalitarian society and reverse dominance hierarchy. *Current Anthropology* 34:227–54.

———. 1996. Emergency decisions, cultural selection mechanics and group selection. *Current Anthropology* 37:763–93.

———. 1999. *Hierarchy in the forest: Egalitarianism and the evolution of human altruism*. Cambridge, Mass: Harvard University Press.

Boerlijst, M. C., M. A. Nowak, and K. Sigmund. 1997. The logic of contrition. *Journal of Theoretical Biology* 185:281–93.

Bowles, S., and H. Gintis. 1998. Is equality passe? *Boston Review* 23, no. 6:4–26.

Boyd, R. 1989. Mistakes allow evolutionary stability in the repeated prisoner's dilemma game. *Journal of Theoretical Biology* 136:47–56.

Boyd, R., and P. J. Richerson. 1985. *Culture and the evolutionary process*. Chicago: University of Chicago Press.

———. 1989. The evolution of indirect reciprocity. *Social Networks* 11:213–36.

———. 1992. Punishment allows the evolution of cooperation (or anything else) in sizable groups. *Ethology and Sociobiology* 13:171–95.

Boyer, P. 1994. *The naturalness of religious ideas: A cognitive theory of religion*. Berkeley: University of California Press.

———. 2001. *Religion explained*. New York: Basic Books.

Bradie, M. 1986. Assessing evolutionary epistemology. *Biology and Philosophy* 1: 401–60.

Bradley, D. E. 1995. Religious involvement and social resources: Evidence from the data set "Americans' Changing Lives." *Journal for the Scientific Study of Religion* 34:259–67.

Buckser, A. 1995. Religion and the supernatural on a Danish island: Rewards, compensators, and the meaning of religion. *Journal for the scientific study of religion* 34:1–16.

Burkert, W. 1996. *Creation of the sacred: Tracks of biology in early religions*. Cambridge, Mass.: Harvard University Press.

Burn, A. R. 1953. Hic breve vivitur. *Past and Present* 4:2–31.

Buss, D. M. 1994. *The evolution of desire: Strategies of human mating*. New York: Basic Books.

———. 1999. *Evolutionary psychology: New science of the mind*. Boston: Allyn and Bacon.

Butler, J. [1726] 1950. *Five sermons*. Indianapolis: Bobbs-Merrill.

Campbell, D. T. 1960. Blind variation and selective retention in creative thought and other knowledge processes. *Psychological Review* 67:380–400.

Cartwright, J. 2000. *Evolution and human behavior.* Cambridge, Mass.: MIT Press.

Cavalli-Sforza, L. L., and D. Carmelli. 1977. The Ashkenazi gene pool: Interpretations. In *Genetic diseases among Ashkenazi Jews,* ed. R. M. Goodman and A. G. Motulsky. New York: Raven Press.

Cavalli-Sforza, L. L., and M. W. Feldman. 1981. *Cultural transmission and evolution.* Princeton, N.J.: Princeton University Press.

Chisholm, J. S. 1999. *Death, hope, and sex.* Cambridge: Cambridge University Press.

Cialdini, R. B., and M. R. Trost. 1998. Social influence: Social norms, conformity, and compliance. In *Handbook of social psychology,* ed. D. T. Gilbert and S. T. Fiske, 151–92. New York: McGraw Hill.

Cohan, F. M. 1984. Can uniform selection retard random genetic divergence between isolated conspecific populations? *Evolution* 38:495–504.

Cohen, G. A. [1978] 1994. *Karl Marx's theory of history: A defense.* Princeton: Princeton University Press. Pages 278–96 reprinted in Martin and McIntyre 1994, 391–402.

Cohn, N. 1966. *Warrant for genocide: The Jewish world conspiracy and the "Protocols of the Elders of Zion."* New York: Harper and Row.

Collcutt, M. 1981. *Five mountains: The Rinzai Zen monastic tradition in Medieval Japan.* Cambridge, Mass.: Harvard University Press.

Collins, R. W. 1968. *Calvin and the libertines of Geneva.* Toronto: Clarke, Irwin, and Co.

Cosmides, L., and J. Tooby. 1992. Cognitive adaptations for social exchange. In *The adapted mind: Evolutionary psychology and the generation of culture,* ed. J. Barkow, L. Cosmides, and J. Tooby, 163–225. New York: Academic Press.

———. 2001. *What is evolutionary psychology? Explaining the new science of the mind.* New Haven: Yale University Press.

Coyne, J. A., N. H. Barton, and M. Turelli. 1997. A critique of Sewall Wright's shifting balance theory of evolution. *Evolution* 51:6643–71.

———. 2000. Is Wright's shifting balance theory important in evolution? *Evolution* 54:306–17.

Csikszentmihalyi, M. 1990. *Flow: The psychology of optimal experience.* New York: Harper and Row.

Cummins, D. D., ed. 1998. *The evolution of mind.* Oxford: Oxford University Press.

Daly, M., and M. Wilson. 1988. *Homicide.* New York: Aldine de Gruyter.

Damasio, A. R. 1994. *Descartes' Error.* New York: Avon.

Darwin, C. 1871. *The descent of man and selection in relation to sex.* New York: Appleton.

———. [1887] 1958. *The autobiography of Charles Darwin, 1809–1882. With original omissions restored.* Edited by Nora Barlow. New York: Harcourt Brace.

Dawkins, R. 1976. *The selfish gene.* 1st ed. Oxford: Oxford University Press.

———. 1982. *The extended phenotype.* Oxford: Oxford University Press.

———. 1998. The emptiness of theology. *Free Inquiry* 18:6.

Dawkins, R., and J. R. Krebs. 1978. Animal signals: Information or manipulation? In *Behavioral ecology: An evolutionary approach,* ed. J. R. Krebs and N. B. Davis, 282–309. Oxford: Blackwell.

Deacon, T. W. 1998. *The symbolic species.* New York: Norton.

Dennett, D. C. 1981. *Brainstorms.* Cambridge, Mass: MIT Press, Bradford Books.

Depew, D. J., and B. H. Weber. 1995. *Darwinism evolving.* Cambridge, Mass: MIT Press.

Detrain, C., J. L. Deneubourg, and J. M. Pasteels, eds. 1999. *Information processing in social insects.* Basel: Birkhauser.

Dobzhansky, T. 1973. Nothing in biology makes sense except in the light of evolution. *American Biology Teacher* 35:125–29.

Du Chaillu, P. B. 1868. *Explorations and adventures in equatorial Africa.* New York: Harper.

Dugatkin, L. A. 1997. *Cooperation among animals.* Oxford: Oxford University Press.

———. 1999. *Cheating monkeys and citizen bees.* New York: Free Press.

———, ed. 1998. *Game theory and animal behavior.* Oxford: Oxford University Press.

Dugatkin, L. A., and M. Alfieri. 1991a. Tit-for-tat in guppies (Poecilia reticulata): The relative nature of cooperation and defection during predator inspection. *Evolutionary Ecology* 5:300–309.

———. 1991b. Guppies and the tit for tat strategy: preference based on past interaction. *Behavioral Ecology and Sociobiology* 28:243–46.

Dunbar, R. I. M. 1996. *Grooming, gossip, and the evolution of language.* Cambridge, Mass: Harvard University Press.

Dunbar, R. I. M., C. Knight, and C. Power, eds. 1999. *The evolution of culture.* New Brunswick, N.J.: Rutgers University Press.

Durham, M. E. 1928. *Some tribal origins, laws, and customs of the Balkans.* London: George Allen and Unwin.

Durham, W. H. 1991. *Coevolution: Genes, culture, and human diversity.* Stanford: Stanford University Press.

Durkheim, E. [1912] 1995. *The elementary forms of religious life.* Translated by Karen E. Fields. New York: The Free Press.

Edelman, G. M. 1988. *Neural darwinism: The theory of neuronal group selection.* New York: Basic Books.

Edelman, G. M., and G. Tonomi. 2001. *A universe of consciousness: How matter becomes imagination.* New York: Basic Books.

Ehrenpreis, A. [1650] 1978. *An epistle on brotherly community as the highest command of love.* In *Brotherly community—the highest command of love: Two Anabap-*

tist documents of 1650 and 1650, by Andreas Ehrenpreis and Claus Felbinger, ed. Robert Friedmann, 9–77. Rifton, N.Y.: Plough Publishing Co.

Eliade, M., ed. 1987. *The encyclopedia of world religions.* New York: Macmillan.

Ellickson, R. C. 1991. *Order without law.* Cambridge, Mass: Harvard University Press.

Ellison, C. G., and L. K. George. 1994. Religious involvement, social ties, and social support in a southeastern community. *Journal for the Scientific Study of Religion* 33:46–61.

Elster, J. 1983. *Explaining technical change.* Cambridge: Cambridge University Press.

Emlen, S. 1975. Migration: Orientation and navigation. In *Avian Biology,* ed. D. Farner and J. King, 129–219. New York: Academic Press.

Endler, J. A. 1986. *Natural selection in the wild.* Princeton: Princeton University Press.

———. 1995. Multiple-trait coevolution and environmental gradients in guppies. *Trends in Ecology and Evolution* 10:22–29.

Evans-Pritchard, E. E. 1940. *The Nuer: A description of the modes of livelihood and political institutions of a Nilotic people.* Oxford: Oxford University Press.

———. 1956. *Nuer religion.* Oxford: Oxford University Press.

———. 1965. *Theories of primitive religion.* Oxford: Clarendon Press.

Faia, M. A. 1986. *Dynamic functionalism.* Cambridge: Cambridge University Press.

Finke, R., and R. Stark. 1992. *The churching of America, 1776–1990.* New Brunswick, N.J.: Rutgers University Press.

Finkelstein, L. 1924. *Jewish self-government in the Middle Ages.* Westport, Conn.: Greenwood Press.

Fischer, D. H. 1989. *Albion's seed: Four British folkways in America.* New York: Oxford University Press.

Frank, R. H. 1988. *Passions within reason.* New York: W. W. Norton.

Frank, S. A. 1998. *Foundations of social evolution.* Princeton: Princeton University Press.

Frazer, J. G. 1890. *The golden bough.* London and New York: Macmillan.

Fukuyama, F. 1995. *Trust.* New York: The Free Press.

Gadagkar, R. 1997. *Survival strategies.* Cambridge, Mass: Harvard University Press.

Gaulin, S. J. C., and D. H. McBurney. 2001. *Psychology: An evolutionary approach.* Upper Saddle River, N.J.: Prentice Hall.

Gigerenzer, G., and U. Hoffrage. 1995. How to improve Bayesian reasoning without instruction. *Psychological Review* 102:684–704.

Godin, J.-G., and L. A. Dugatkin. 1996. Female mating preference for bold males in the guppy. *Proceedings of the National Academy of Sciences, USA* 93:10262–67.

Goodenough, U. 1998. *The sacred depths of nature.* Oxford: Oxford University Press.

Gould, S. J. 1989. *Wonderful life*. New York: W. W. Norton.

———. 1999. *Rock of ages: Science and religion in the fullness of life*. New York: Ballantine.

Gould, S. J., and R. C. Lewontin. 1979. The spandrels of San Marco and the panglossian paradigm: A critique of the adaptationist program. *Proceedings of the Royal Society of London*, Series B, *Biological Sciences* 205:581–98.

Guthrie, S. E. 1995. *Faces in the clouds: A new theory of religion*. Oxford: Oxford University Press.

Hamilton, W. D. 1964. The genetical evolution of social behavior: I and II. *Journal of Theoretical Biology* 7:1–52.

———. 1975. Innate social aptitudes in man: An approach from evolutionary genetics. In *Biosocial anthropology*, ed. R. Fox, 133–53. London: Malaby Press.

———. 1996. *The narrow roads of gene land*. Oxford: W. H. Freeman/Spektrum.

Hammer, M. F., A. J. Redd, E. T. Wood, M. R. Bonner, H. Jaranazi, T. Korafet, S. Sanachiora-Benerocetti, A. Oppenheim, M. A. Jobling, J. Jenkins, H. Ostrer, and B. Bonnie-Tamir. 2000. Jewish and Middle Eastern non-Jewish populations share a common pool of Y-chromosome biallelic haplotypes. *Proceedings of the National Academy of Sciences, USA* 97:6769–74.

Hart, K. E. 1999. A spiritual interpretation of the 12-steps of Alcoholics Anonymous: From resentment to forgiveness to love. *Journal of Ministry in Addiction and Recovery* 6:25–39.

Hartung, J. 1995a. Love thy neighbor: The anatomy of in-group morality. *Skeptic* 3 (November): 86–99.

———. 1995b. Review of "A people that shall dwell alone: Judaism as a group evolutionary strategy." *Ethology and Sociobiology* 16:335–42.

Harvey, W. [1628] 1995. *The anatomical excercises*. New York: Dover.

Hauser, M., and S. Carey. 1998. Building a cognitive creature from a set of primitives: Evolutionary and developmental insights. In *The evolution of mind*, ed. D. D. Cummins, 51–106. Oxford: Oxford University Press.

Hayek, F. 1988. *The fatal conceit*. London: Routledge.

Heckathorn, D. D. 1990. Collective sanctions and compliance norms: A formal theory of group-mediated social control. *American Sociological Review* 55:366–84.

———. 1993. Collective action and group heterogeneity: Voluntary provision vs. selective incentives. *American Sociological Review* 58:329–50.

Hempel, C. G. 1959. The logic of functional analysis. In *Symposium on sociological theory*, ed. L. Gross. New York: Harper and Row. Reprinted in Martin and McIntyre 1994, 349–76.

Herre, E. A. 1999. Laws governing species interactions? Encouragement and caution from figs and their associates. In *Levels of selection in evolution*, ed. L. Keller, 209–37. Princeton: Princeton University Press.

Hesselink, I. J. 1997. *Calvin's first catechism: A commentary.* Louisville, Ky.: Westminster John Knox Press.

Heyes, C., and L. Huber, eds. 2000. *The evolution of cognition.* Cambridge, Mass.: MIT Press.

Hinde, R. 1999. *Why gods persist: A scientific approach to religion.* New Brunswick, N.J.: Routledge.

Hirschi, T., and R. Stark. 1969. Hellfire and delinquency. *Social Problems* 17:202–13.

Hirshleifer, J. 1977. Economics from a biological viewpoint. *Journal of Law and Economics* 20:1–52.

Hodgson, G. M. 1993. *Economics and evolution.* Cambridge, England: Polity Press.

Hoge, D. R., and D. A. Roozen. 1979. *Technical appendix to understanding church growth and decline, 1950–1978.* Hartford, Conn.: Hartford Seminary Foundation.

Houde, A. E., and M. Hankes. 1997. Mismatch between female preference and expression of male color pattern in some guppy populations. *Animal Behavior* 53:353–61.

Howell, S. 1984. *Society and cosmos: Chewong of peninsular Malaya.* Singapore: Oxford University Press.

Hull, D. 1973. *Darwin and his critics.* Chicago: University of Chicago Press.

Huxley, T. 1863. *Evidence as to man's place in nature.* London: Williams and Norgate.

Iannaccone, L. R. 1992. Sacrifice and stigma: Reducing free-riding in cults, communes, and other collectives. *Journal of Political Economy* 100:271–91.

———. 1994. Why strict churches are strong. *American Journal of Sociology* 99: 1180–211.

———. 1995. Voodoo economics? Reviewing the rational choice approach to religion. *Journal for the Scientific Study of Religion* 34:76–89.

Ingle, H. L. 1994. *First among friends: George Fox and the creation of Quakerism.* New York: Clarendon.

Jacob, F. 1977. Evolution and tinkering. *Science* 196:1161–66.

James, W. 1902. *The varieties of religious experience.* New York: Longmans, Green.

Johnson, P. 1976. *A history of Christianity.* New York: Atheneum.

Kaplan, H., and K. Hill. 1985a. Hunting ability and reproductive success among male Ache foragers. *Current Anthropology* 26:131–33.

———. 1985b. Food sharing among Ache foragers: Tests of explanatory hypotheses. *Current Anthropology* 26:223–45.

Kaplan, H., K. Hill, and A. Hurtado. 1984. Food sharing among the Ache hunter-gatherers of eastern Paraguay. *Current Anthropology* 25:113–15.

Katz, J. 1961. *Tradition and crisis: Jewish society at the end of the middle ages.* New York: Free Press of Glencoe.

Kelly, D. M. 1972. *Why conservative churches are growing: A study in the sociology of religion*. Macon, Ga.: Mercer University Press.

Kelly, R. C. 1985. *The Nuer conquest*. Ann Arbor: University of Michigan Press.

Kincaid, H. 1990. Assessing functional explanation in the social sciences. In *PSA 90*, ed. A. Fine, M. Forbes, and L. Wessels. Reprinted in Martin and MacIntyre 1994, 415–28.

Klir, G., U. St. Clair, and B. Juan. 1997. *Fuzzy set theory: Foundations and applications*. New York: Prentice Hall.

Knauft, B. M. 1991. Violence and sociality in human evolution. *Current Anthropology* 32:391–428.

Kobyliansky, E., S. Micle, M. Goldschmidt-Nathan, B. Arensburg, and H. Nathan. 1982. Jewish populations of the world: Genetic likenesses and differences. *Annals of Human Biology* 9:1–34.

Konner, M. 1999. Darwin's truth, Jefferson's vision. *American Prospect* 45:30–38.

Kosko, B. 1993. *Fuzzy thinking*. New York: Hyperion.

Kwilecki, S., and L. S. Wilson. 1998. Was Mother Teresa maximizing her utility? An idiographic application of rational choice theory. *Journal for the Scientific Study of Religion* 37:205–21.

Kwon, V. H., H. R. Ebaugh, and J. Hagan. 1997. The structure and functions of cell group ministry in a Korean Christian church. *Journal for the Scientific Study of Religion* 36:247–56.

Lack, D. 1961. *Darwin's finches*. Cambridge: Cambridge University Press.

Laland, K. N., J. Odling-Smee, and M. W. Feldman. 2000. Niche construction, biological evolution, and cultural change. *Biological and Brain Sciences* 23:131–75.

Landa, J. T. 2000. The law and bioeconomics of ethnic cooperation and conflict in plural societies of Southeast Asia: A theory of Chinese merchant success. *Journal of Bioeconomics* 1:269–84.

Lansing, J. S. 1991. *Priests and programmers: Technologies of power in the engineered landscape of Bali*. Princeton: Princeton University Press.

———. 2000. Anti-chaos, common property, and the emergence of cooperation. In *Dynamics in human and primate societies*, ed. T. A. Kohler and G. J. Gumerman, 207–24. Oxford: Oxford University Press.

Lessnoff, M. H. 1994. *The spirit of capitalism and the Protestant ethic: An enquiry into the Weber thesis*. London: Aldershot.

Levine, R. A., and D. T. Campbell. 1972. *Ethnocentrism: Theories of conflict, ethnic attitudes, and group behavior*. New York: Wiley.

Lumsden, C. J., and E. O. Wilson. 1981. *Genes, mind, and culture*. Cambridge, Mass.: Harvard University Press.

MacDonald, K. B. 1994. *A people that shall dwell alone: Judaism as a group evolutionary strategy*. Westport, Conn.: Praeger.

———. 1998a. *Separation and its discontents: Toward an evolutionary theory of anti-Semitism*. Westport, Conn.: Praeger.

———. 1998b. *The culture of critique: An evolutionary analysis of Jewish involvement in twentieth-century intellectual and political movements.* Westport, Conn.: Praeger.

Malinowski, B. 1944. *A scientific theory of culture.* Chapel Hill: University of North Carolina Press.

———. 1948. *Magic, science, and religion and other essays.* Boston: Beacon.

Margulis, L. 1970. *Origin of eukaryotic cells.* New Haven: Yale University Press.

Martin, M., and L. C. McIntyre, eds. 1994. *Readings in the philosophy of social science.* Cambridge, Mass.: MIT Press.

Maynard Smith, J. 1982. *Evolution and the theory of games.* Cambridge: Cambridge University Press.

Maynard Smith, J., and E. Szathmary. 1995. *The major transitions of life.* New York: W. H. Freeman.

McCullough, M. E., K. I. Pargament, and C. E. Thoresen, eds. 2000. *Forgiveness: Theory, research, and practice.* New York: Guilford Press.

McGrath, A. E. 1990. *A life of John Calvin.* Oxford: Blackwell.

McNeill, J. T. 1954. *The history and character of Calvinism.* Oxford: Oxford University Press.

McNeill, W. H. 1976. *Plagues and peoples.* Garden City, N.Y.: Doubleday.

Meeks, W. A., and R. L. Wilken. 1978. *Jews and Christians in Antioch in the first four centuries of the common era.* Missoula, Mont.: Scholar's Press.

Michod, R. E. 1999a. *Darwinian dynamics.* Princeton: Princeton University Press.

———. 1999b. Individuality, immortality, and sex. In *Levels of selection in evolution,* ed. L. Keller, 53–74. Princeton: Princeton University Press.

Miller, K. R. 1999. *Finding Darwin's God: A scientist's search for common ground between God and evolution.* New York: HarperCollins.

Monter, E. W. 1967. *Calvin's Geneva.* New York: John Wiley and Sons.

Naphy, W. G. 1994. *Calvin and the consolidation of the Genevan reformation.* New York: St. Martin's Press.

Nisbett, R. E., and D. Cohen. 1996. *Culture of honor.* New York: Westview Press.

Olson, J. E. 1989. *Calvin and social welfare.* London: Selinsgrove.

Ostrom, E. 1991. *Governing the commons: The evolution of institutions of collective action.* Cambridge: Cambridge University Press.

Ostrom, E., R. Gardner, and J. M. Walker. 1994. *Rules, games, and common-pool resources.* Ann Arbor: University of Michigan Press.

Pagels, E. 1995. *The origin of Satan.* Princeton: Princeton University Press.

Paley, W. 1805. *Natural theology.* London: Rivington.

Payne, J. W., J. R. Bettman, and E. J. Johnson. 1993. *The adaptive decision maker.* Cambridge: Cambridge University Press.

Phillips, K. P. 2000. *The cousins' war: Religion, politics, and the triumph of Anglo-America.* New York: Basic Books.

Plotkin, H. 1994. *Darwin machines and the nature of knowledge*. Cambridge, Mass.: Harvard University Press.

Pomiankowski, A. 1999. Intragenomic conflict. In *Levels of selection in evolution*, ed. L. Keller, 121–52. Princeton: Princeton University Press.

Prager, D., and J. Telushkin. 1983. *Why the Jews? The reason for antisemitism*. New York: Simon and Schuster.

Provine, W. B. 1986. *Sewall Wright and evolutionary biology*. Chicago: University of Chicago Press.

Putnam, R. D. 1992. *Making democracy work: Civic traditions in modern Italy*. Princeton: Princeton University Press.

Ragin, C. C. 2000. *Fuzzy-set social science*. Chicago: University of Chicago Press.

Rapoport, A. 1991. Ideological commitments and evolutionary theory. *Journal of Social Issues* 47:83–100.

Rappaport, R. 1979. *Ecology, meaning, and religion*. Richmond, Calif.: North Atlantic Books.

Rawls, J. 1971. *A theory of justice*. Cambridge Mass.: Harvard University Press.

Reid, J. K. S. 1954. *Calvin: Theological treatises*. Philadelphia: Westminster Press.

Richerson, P. J., and R. Boyd. 1999. Complex societies: The evolutionary origins of a crude superorganism. *Human Nature* 10:253–90.

———. 2000. Climate, culture, and the evolution of cognition. In *The evolution of cognition*, ed. C. Heyes and L. Huber, 329–46. Cambridge, Mass.: MIT Press.

Rolls, E. 1998. *The brain and emotion*. Oxford: Oxford University Press.

Ruse, M. 2000. *Can a Darwinian be a Christian? The relationship between religion and science*. Cambridge: Cambridge University Press.

Russell, J. C. 1958. Late ancient and medieval population. *Transactions of the American Philosophical Society* n.s. 48, pt. 3.

Sahlins, M. D. 1976. *The use and abuse of biology: An anthropological critique of sociobiology*. Ann Arbor: University of Michigan Press.

Scher, S. J., and F. Rauscher, eds. Forthcoming. *Evolutionary psychology: Alternative approaches*. Dordrecht: Kluwer Academic Publishers.

Schwartz, J. 2000. Death of an altruist. *Lingua Franca* 10, no. 5:51–61.

Seeley, T. D. 1995. *The wisdom of the hive: The social psychology of honey bee colonies*. Cambridge, Mass.: Harvard University Press.

Seeley, T. D., and S. C. Buhrman. 1999. Group decision making in swarms of honey bees. *Behavioral Ecology and Sociobiology* 45:19–31.

Sherif, M., L. J. Harvey, B. J. White, W. R. Hood, and C. W. Sherif. [1961] 1988. *The robber's cave experiment: Intergroup conflict and cooperation*. Middletown, Conn.: Wesleyan University Press.

Simon, H. A. 1990. A mechanism for social selection and successful altruism. *Science* 250:1665–68.

Singer, I. B. 1962. *The Slave*. Translated by I. B. Singer and C. Hemley. New York: Farrar, Straus and Cudahy.

Skyrms, B. 1996. *Evolution of the social contract.* Cambridge: Cambridge University Press.

Smith, H. [1958] 1991. *The world's religions.* San Francisco: HarperSanFrancisco.

Sober, E. 1999. The multiple realizability argument against reductionism. *Philosophy of Science* 66:542–64.

———. 2001. The design argument. In *Blackwell companion to the philosophy of religion,* ed. W. Mann. New York: Blackwell.

Sober, E., and D. S. Wilson. 1998. *Unto others: The evolution and psychology of unselfish behavior.* Cambridge, Mass.: Harvard University Press.

Somé, M. P. 1993. *Ritual: power, healing, and community.* New York: Penguin.

Sperber, D. 1996. *Explaining culture.* Cambridge, Mass.: Blackwell.

Stark, R. 1996. *The rise of Christianity: A sociologist reconsiders history.* Princeton: Princeton University Press.

———. 1999. Micro foundations of religion: A revised theory. *Sociological Theory* 17:264–89.

Stark, R., and W. S. Bainbridge. 1985. *The future of religion.* Berkeley: University of California Press.

———. 1987. *A theory of religion.* New Brunswick, N.J.: Rutgers University Press.

———. 1997. *Religion, deviance, and social control.* New York: Routledge.

Stark, R., and R. Finke. 2000. *Acts of faith.* Berkeley: University of California Press.

Sterelny, K., and P. E. Griffiths. 1999. *Sex and death: An introduction to philosophy of biology.* Chicago: University of Chicago Press.

Stevens, A., and J. Price. 1996. *Evolutionary psychiatry: A new beginning.* London: Routledge.

Swanson, G. E. 1960. *The birth of the gods.* Ann Arbor: University of Michigan Press.

———. 1967. *Religion and regime: A sociological account of the Reformation.* Ann Arbor: University of Michigan Press.

Swenson, W., J. Arendt, and D. S. Wilson. 2000. Artificial selection of microbial ecosystems for 3-chloroaniline biodegradation. *Environmental Microbiology* 2: 564–71.

Swenson, W., D. S. Wilson, and R. Elias. 2000. Artificial Ecosystem Selection. *Proceedings of the National Academy of Sciences* 97:9110–14.

Tajfel, H. 1981. *Human groups and social categories: Studies in social psychology.* Cambridge: Cambridge University Press.

Thornhill, R. 1998. Darwinian aesthetics. In *Handbook of Evolutionary Psychology,* ed. C. Crawford and D. Krebs, 543–72. Mahway, N.J.: L. Erlbaum.

Tocqueville, A. d. [1835] 1990. *Democracy in America.* Translated by G. Lawrence. Garden City, N.Y.: Anchor Books.

Tomasello, M. 1999. *The cultural origins of human cognition.* Cambridge, Mass.: Harvard University Press.

Tooby, J., and L. Cosmides. 1992. The psychological foundations of culture. In

The adapted mind: Evolutionary psychology and the generation of culture, ed. J. H. Barkow, L. Cosmides, and J. Tooby, 19–136. Oxford: Oxford University Press.

Turnbull, C. M. 1965. *The Mbuti Pygmies: An ethnographic survey.* New York: American Museum of Natural History.

Turner, V. [1969] 1995. *The ritual process: Structure and anti-structure.* New York: Aldine de Gruyter.

Tylor, E. B. 1871. *Primitive culture: Research into the development of mythology, philosophy, religion, art, and custom.* London: J. Murray.

Van Schaik, C., and C. H. Janson, eds. 2000. *Infanticide by males and its implications.* Cambridge: Cambridge University Press.

Wade, M. J., and C. J. Goodnight. 1998. The theories of Fisher and Wright in the context of metapopulations: When nature does many small experiments. *Evolution* 52:1537–53.

Wallace, R. S. 1988. *Calvin, Geneva, and the reformation.* Edinburgh: Scottish Academic Press.

Watkins, J. W. N. 1957. Historical explanations in the social sciences. *British Journal for the Philosophy of Science* 8:104–17.

Weber, M. [1930] 1992. *The Protestant ethic and the spirit of capitalism.* Translated by Talcott Parsons. London: Routledge.

Wegner, D. M. 1986. Transactive memory: A contemporary analysis of the group mind. In *Theories of group behavior,* ed. B. Mullen and G. R. Goethals, 185–208. New York: Springer-Verlag.

Wehner, R., and M. Srinivasan. 1981. Searching behavior in desert ants, genus Cataglyphis (Formicidae, Hymenoptera). *Journal of Comparative Physiology,* Series A, *Sensory, Neural, and Behavioral Physiology* 142:315–38.

Weiner, J. 1994. *The beak of the finch: A story of evolution in our time.* New York: Knopf.

Wesley, J. 1976. Thoughts upon Methodism. In *The works of John Wesley,* vol. 9: *The Methodist societies.* Edited by Rupert E. Davies. Nashville: Abingdon Press.

Wheeler, W. M. 1928. *The social insects: Their origin and evolution.* New York: Harcourt, Brace.

Williams, G. C. 1966. *Adaptation and natural selection: A critique of some current evolutionary thought.* Princeton: Princeton University Press.

———. 1992. *Natural selection: Domains, levels and challenges.* Oxford: Oxford University Press.

———. 1996. *Plan and purpose in nature.* London: Weidenfeld and Nicolson.

Wilson, D. S. 1975. A theory of group selection. *Proceedings of the National Academy of Sciences* 72:143–46.

———. 1988. Holism and reductionism in evolutionary biology. *Oikos* 53:269–73.

———. 1990. Species of thought: A comment on evolutionary epistemology. *Biology and Philosophy* 5:37–62.

———. 1992. The effect of complex interactions on variation between units of a metacommunity, with implications for biodiversity and higher levels of selection. *Ecology* 73:1984–2000.

———. 1994. Adaptive genetic variation and human evolutionary psychology. *Ethology and Sociobiology* 15:219–35.

———. 1995. Language as a community of interacting belief systems: A case study involving conduct toward self and others. *Biology and Philosophy* 10:77–97.

———. 1997. Incorporating group selection into the adaptationist program: A case study involving human decision making. In *Evolutionary social psychology*, ed. J. Simpson and D. Kendricks, 345–86. Mahwah, N.J.: Erlbaum.

———. 1998a. Hunting, sharing and multilevel selection: The tolerated theft model revisited. *Current Anthropology* 39:73–97.

———. 1998b. Evolutionary game theory and human behavior. In *Game theory and animal behavior*, ed. L. A. Dugatkin and H. K. Reeve, 261–82. Oxford: Oxford University Press.

———. 1998c. Adaptive individual differences within single populations. *Philosophical Transactions of the Royal Society of London*, Series B, *Biological Sciences* 353: 199–205.

———. 1999a. A critique of R. D. Alexander's views on group selection. *Biology and Philosophy* 14:431–49.

———. 1999b. Tasty slice—but where is the rest of the pie? Review of *Evolutionary psychology*, by David Buss. *Evolution and Human Behavior* 20:279–87.

———. 2000. Nonzero and nonsense: Group selection, nonzerosumness, and the human Gaia hypothesis. *Skeptic* 8:84–89.

———. Forthcoming. Evolution, morality and human potential. In *Evolutionary psychology: Alternative approaches*, ed. S. J. Scher and F. Rauscher. Dordrecht: Kluwer Academic Publishers.

Wilson, D. S., A. B. Clark, K. Coleman, and T. Dearstyne. 1994. Shyness and boldness in humans and other animals. *Trends in Ecology and Evolution* 9:442–46.

Wilson, D. S., and L. A. Dugatkin. 1997. Group selection and assortative interactions. *American Naturalist* 149:336–51.

Wilson, D. S., and K. M. Kniffin. 1999. Multilevel selection and the social transmission of behavior. *Human Nature* 10:291–310.

Wilson, D. S., D. Near, and R. R. Miller. 1996. Machiavellianism: A synthesis of the evolutionary and psychological literatures. *Psychological Bulletin* 199:285–99.

Wilson, D. S., and E. Sober. 2001. The fall and rise and fall and rise and fall and rise of altruism in evolutionary biology. In *Altruism and altruistic love: Science, philosophy, and religion in dialogue*, ed. S. G. Post, L. G. Underwood, J. P. Schloss, and W. B. Hurlburt. Oxford: Oxford University Press.

Wilson, D. S., and J. Yoshimura. 1994. On the coexistence of specialists and generalists. *American Naturalist* 144:692–707.

Wilson, E. O. 1998. *Consilience.* New York: Knopf.

Wrangham, R., and D. Peterson. 1997. *Demonic males: Apes and the origin of human violence.* New York: Houghton Mifflin.

Wright, R. 2000. *Nonzero: The logic of human destiny.* New York: Pantheon.

Wynne-Edwards, V. C. 1962. *Animal dispersion in relation to social behaviour.* Edinburgh: Oliver and Boyd.

ACKNOWLEDGMENTS

One of the joys of studying evolution is that it provides a passport to so many subjects. The same theory that explains the pageant of life on earth also helps to explain the pageant of human life. Throughout my career I have used my passport to study organisms as diverse as microbes, zooplankton, mites, insects, amphibians, fish, birds—and people. With respect to people I have studied topics as diverse as individual differences in personality traits, gossip as much more than "small talk," physical beauty as much more than a physical characteristic—and now religion.

This nomadic approach to science would be impossible without the help of fellow travelers and colleagues more deeply rooted in the study of single subjects. I know I have learned from them and only hope that I have given something in return. Two fellow travelers deserve special mention with respect to this book; philosopher Elliott Sober and anthropologist Chris Boehm. Philosophy gave birth to the sciences, and Elliott Sober still tends over them with his insightful analyses of conceptual issues. I am pleased and flattered to be his associate. Chris Boehm is equally distinguished as a cultural anthropologist and a primatologist. His ability to combine these bodies of knowledge in a way that does justice to both should serve as a role model for the future.

It is hard to describe the pleasure of landing on the shore of a new subject such as religion, armed with nothing but a theoretical road map. Everything is simultaneously exotic and familiar. Almost everyone I encountered was generous with their time and effort. John Flint contributed to my early religious education and pointed me to several books that proved foundational. Diego Gambetta served as my guide for the subject of functionalism in the social sciences. I discovered how good scientific research on religion can be in the work of Rodney Stark, first by reading his books and then by interacting with the person. The fact that we disagree on some issues does not detract from my admiration.

I was lucky to receive a grant from the John Templeton foundation to finance my expedition. In his own books, Sir John Templeton emphasizes the practical wisdom of religion and spirituality, which science can help to enlighten rather than to oppose. I hope this book contributes to the synthesis that he has in mind. I also want to thank members of the staff of the John Templeton Foundation, whom I have grown to know and enjoy during the conferences they have organized.

Many colleagues have provided feedback on the material in this book, including John Avise, Chris Boehm, Anne Clark, Eric Dietrich, Lee Dugatkin, John Flint, Frank Foreman, Francis Fukuyama, Herb Gintis, Loren Haarsma, Kevin Kniffin, Steve Lansing, Kevin MacDonald, Becka Moldover, Rick O'Gorman, Robert Richards, Peter Richerson, Martin Riesebrodt, Paul Roberts, Michael Ruse, Robert Sapolsky, Jeff Schloss, Tom Seeley, Elliott Sober, Kim Sterelny, Lisa Strick, Nicholas Thompson, Katie Wilson, and Todd Zwicki. My sister, Becka Moldover, provided an especially valuable service by helping me stick to the basics. Christie Henry is a consummate editor, and the University of Chicago Press has been a pleasure to work with. Richard Allen improved the manuscript not only as a copy editor but as a religious scholar. Needless to say, the residual mistakes are all mine. John Calvin once ended a letter with the words "God keep you and show you how much the blows of a sincere friend are better than the treacherous flattery of certain others." I have many sincere friends!

Finally, I would like to thank my wife and colleague, Anne B. Clark, and my two daughters, Tamar and Katie. Most pilgrims must travel alone but I get to travel in the company of a warm, exuberant, and intellectually stimulating family. I dedicate this book to them.

INDEX

Abrams, D., 27, 139
adaptation, 6–8, 44–45, 67–68, 70–73, 95, 122–123, 187, 235; studying at appropriate scale, 91–92, 177–182, 188
adaptationist program, 70–73, 171–177, 187–188, 239, 243
adaptive flexibility. *See* phenotypic plasticity
adaptive landscape, 39, 238–239
afterlife, 60, 155–156, 186
alcoholism, 140, 179–181
Alexander, R. D., 12, 28, 44, 100
Alfieri, M., 193
Allen, N. J., 66
Almann, J. M., 75, 77
altruism, 2, 8, 11–12, 18–20, 24–25, 154–155, 176, 189–194, 211, 243
Anabaptists, 89
animism, 52
Antioch, 149–151
anti-Semitism, 138–139, 142–143, 144–147, 216, 242
ants, 26
Arendt, J., 239
Arensburg, B., 137
argument from design, 71–72, 115–119, 240
artificial intelligence, 28
Australian aborigines, 56

averaging fallacy, 13–14, 18, 236, 237
Axelrod, R., 190, 193, 236

bacteria, 75
Bainbridge, W. S., 48–49, 82, 179–181, 186–187, 220, 229
Bali. *See* water temple system of Bali
baptism, 210–211
Barkow, J. H., 11, 238
Barton, N. H., 239
beauty, 231–233
belief system, 22–25, 40–43, 98–105
Bettman, J. R., 34
Blackmore, S., 28, 44, 53
Blurton Jones, N., 237
Boak, A. E. R., 152
Boehm, C., 11, 20, 21, 26, 32, 34, 41, 83–84, 200–203, 223–224, 225, 235
Boerlijst, M. C., 192
Bonner, M. R., 137
Bonnie-Tamir, B., 137
boundary maintenance, 81–82
Bowles, S., 224, 225
Boyd, R., 26, 31, 34, 40, 192, 225, 226–227, 238, 239, 244
Boyer, P., 44, 53, 239
Bradie, M., 41
Bradley, D. E., 177–178
Buckser, A., 47, 49

Buhrman, S. C., 34
Burkert, W., 239
Burn, A. R., 155
Buss, D. M., 29–30, 96, 238
Butler, J., 7
byproduct hypotheses, 44–45, 49, 52–53, 79–80, 83, 118, 149, 155, 185

Calvin, J., 88–91, 103, 111–113, 114, 121
Calvinism, 88–124, 130, 176–177, 211, 240; Calvin's catechism, 93–104; decision-making structure, 110; educational system, 110; health and welfare system, 110–111
Campbell, D. T., 122, 135
Carey, S., 6
Carmelli, D., 137
Cartwright, J., 238
Cataglyphis bicolor, 26
catechisms, 93–104
Cavalli-Sforza, L. L., 137, 238
cells, 151; cell group ministry, 167–168; eukaryotic, 17
cheating, 26
Chewong, 22–25, 237
Chicago school of economists, 75–76
Chisholm, J. S., 238
church-sect theory, 182–187
Cialdini, R. B., 174
Civil War, American, 79
Clark, A. B., 74
climatic instability, 31
Cohan, F. M., 67
Cohen, D., 79
Cohen, G. A., 132
Cohn, N., 138
Coleman, K., 74
Collcutt, M., 1
Collins, R. W., 109, 111–112, 113, 117, 135
communitas, 61–63
complexity, 68–69, 239
conflict resolution, 60
conformity, 26, 195–204
consciousness, 32, 77–79, 121

context-sensitivity, 59, 74, 98, 109, 189, 217
contrition, 191–192
cooperation, 17, 27, 140. *See also* altruism
Cop, N., 88
Corinthians, 212
Cosmides, L., 11, 26, 28–29, 85, 238
Coyne, J. A., 239
creationists, 73
cryptic coloration, 8, 68, 73
Csikszentmihalyi, M., 219
cults, 182
cultural evolution, 2, 10–11, 28–35, 77–79, 119–120, 133, 139, 140, 198–204, 219–220, 237–238, 242, 244
cultural parasite, 44
cultural relativism, 132–133, 160

Dagara, 63–65
Daly, M., 96
Damasio, A. R., 77, 174, 193
Darwin, C., 5, 9, 18, 87, 125, 160
Darwin machines, 31–32, 35, 77–78
Dawkins, R., 2, 16, 18, 28, 44, 53, 97, 236–237, 239
Deacon, T. W., 26, 54, 75, 77, 226
dead reckoning, 26
Dearstyne, T., 74
decision-making, 26, 102–103, 110, 115, 129, 240
Deneubourg, J. L., 33
Dennett, D. C., 244
Depew, D. J., 239
design stance, 244
Detrain, C., 33
Deuteronomy, 134–135, 142–143
developmental flexibility. *See* phenotypic plasticity
deviant behavior, 179–182
Diaspora, 139–143
Dinka, 78–79, 137
disease, 149–151, 153–154
Dobzhansky, T., 7
Drosophila, 67
Du Chaillu, P. B., 62
Dugaktin, L. A., 15, 193

Dunbar, R. I. M., 238
Durham, M. E., 200–202
Durham, W. H., 238, 244
Durkheim, E., 47, 52–55, 63, 70, 80, 82, 83, 133, 156, 159, 215, 222, 226

early Christian church, 147–157, 205–216
Ebaugh, H. R., 165–169
Ecclesiastical Ordinances, 106–111
ecological fallacy, 181
economic theory. See rational choice theory
Edelman, G. M., 238
egalitarianism, 21–25, 43, 62–63, 120, 127, 199, 225
Ehrenpreis, A., 1
elders, 107–108, 167
Eliade, M., 126, 157
Elias, R., 239
Ellickson, R. C., 27
Ellison, C. G., 177–178
Elster, J., 74–77, 84
Emlen, S., 26
emotion, 42, 194–195
Endler, J. A., 36, 71, 73
epistemology, 41
Essenes, 206, 208
Evans-Pritchard, E. E., 55–61, 70, 83, 124, 199–200
evolutionary game theory, 16, 189–194, 236
evolutionary psychology, 25–26, 29–30, 237, 238

Faia, M. A., 69
faith, 101–103
Farel, G., 89–90
Feldman, M. W., 238, 240, 244
fighting behavior, 16
Finke, R., 41, 48
Finkelstein, L., 136
firm, theory of , 80, 83
Fischer, D. H., 79
fitness, 37–40, 79

forgiveness, 97–98, 100–101, 189–218, 244; Christian, 204–215
Frank, R. H., 192
Frank, S. A., 235
Frazer, J. G., 41, 44
freeloading, 24, 82, 156, 165, 175
fruit flies. See Drosophila
Fukuyama, F., 204
functionalism, 6–7, 37, 48, 49, 52–55, 65–79, 83–85, 125, 132–133, 157, 164, 220–222, 241; designing agents, 73–79, 121–122; latent vs. manifest functions, 76–79
fundamental problem of social life, 7–8, 9, 18–19, 49, 118, 162–163
fuzzy sets, 220–222

Galapagos Islands, 92
game theory. See evolutionary game theory
Gardner, E., 27
Gaulin, S. J. C., 237, 238
generosity, 192
genetic determinism, 28, 237
Geneva, 89–91, 109–110, 123–124
George, L. K., 177–178
Gigerenzer, G., 30
Gintis, H., 224, 225
Godin, J.-G., 193
Golden Rule, 96–97, 98, 109
Goldschmidt-Nathan, M., 137
Goodenough, U., 239
Goodnight, C. J., 239
Gospels, 207–216
Gould, S. J., 10, 44, 70, 207, 215, 239
green revolution, 129–130
Griffiths, P. E., 236, 238
group mind, 32–34, 77
group selection, 9–10, 16–17, 19–20, 38, 45, 81, 138–140, 235, 240; how to define groups, 14–17, 236; human groups as adaptive units, 20–25, 35–37; and idealized religion, 46; procedure for seeing, 12; rejection of, 11–12, 125. See also multilevel selection
Gruet, J., 113

guppies. *See Peocilia reticulata*
Guthrie, S. E., 44

Hagan, J., 165–169
Hamilton. W. D., 11, 16, 139, 242–243
Hammer, M. F., 137
Hankes, M., 73
Hart, K. E., 181
Hartung, J., 133, 135
Harvey, L. J., 27
Harvey, W., 121
Hauser, M., 6
Hayek, F., 240
Heckathorn, D. D., 19
Hempel, C. G., 70
Hesselink, I. J., 93
Heyes, C., 238
Hill, K., 21
Hinde, R., 239
Hirschi, T., 179
Hirshleifer, J., 173
Hoffrage, U., 30
Hoge, D. R., 163
Hogg, M. A., 27, 139
holism, 66–68, 239
Holocaust, 139
homicide, 199–200
Hood, W. R., 27
Houde, A. E., 73
Howell, S., 22–25, 237
Huber, L., 238
Hull, D., 161
human relations area file (HRAF), 157, 237
humility, 57, 64
hunter-gatherers, 11, 20–25, 42–43, 120, 127, 195–198, 199, 216, 237
Hurtado, A., 21
Hutterites, 1, 46
Huxley, T., 2

Iannaccone, L. R., 81–83, 136, 156, 163–165
imitation, 32, 75–76
immune system, 30–31, 238
inclusive fitness theory. *See* kin selection theory

individual differences, 244
individualism, 6, 85; methodological, 66–68
individual selection, 13, 118, 184. *See also* multilevel selection
infanticide, 38, 151, 152, 243
Ingle, H. L., 185, 209
inheritance, 39
intentional behavior, 73–79
intentional stance, 244
internalization, 61, 103–104
intragenomic conflict, 40, 237
irrigation, 126–133, 241
Islam, 133, 137, 242
Italy, 79

Jacob, F., 120
James, W., 47
Janson, C. H., 38
Jaranazi, H., 137
Jenkins, J., 137
Jero Gde, 126
Jesus, 206–208
Jobling, M. A., 137
John, 215
Johnson, E. J., 34
Johnson, P., 154
Juan, B., 222
Judaeo-Christian belief, 91, 97, 134
Judaism, 57, 60, 133–147, 148, 205–207, 242, 243; cultural isolating mechanisms, 136–138; surviving intergroup interactions, 138–147
Judas, 211
Justin, 208–209, 210–211, 218
Just-so story, 55, 70–73, 116

Kaplan, H., 21
Katz, J., 143
Kelly, D. M., 82
Kelly, R. C., 78–79, 137
Kincaid, H., 69
kin selection theory, 11, 16, 22, 139, 242
Klir, G., 222
Knauft, B. M., 21
Kniffin, K. M., 34, 237, 238, 244
Knight, C., 238

Kobyliansky, E., 137
Konner, M., 11–12, 20
Korafet, T., 137
Korean Christian Church, 165–169, 173, 177, 181, 183–184
Kosko, B., 222
Krebs, J. R., 97
Kwilecki, S., 80
Kwon, V. H., 165–169

Lack, D., 92
Laland, K. N., 240, 244
Landa, J. T., 243
Lansing, J. S., 127–133, 242
leaders, 21
learning theory, 25, 29, 85
leopard-skin chief, 21, 60–61, 199–200
Lessnoff, M. H., 120
Levine, R. A., 135
Leviticus, 134
Lewontin, R. C., 10, 44, 70, 215, 239
life cycle of denominations, 182–187
love, 1, 46
Luke, 208, 213
Lumsden, C. J., 238, 244
Luther, M., 88

Maccabean revolt, 205–206
MacDonald, K. B., 133, 138, 139, 142–143
major transitions of life, 17–20, 22, 25, 86, 222–223
Malinowski, B., 41, 69
Margulis, L., 17
Mark, 214
Martin, M., 66, 69, 70
Marxism, 128, 132, 160
Matthew, 154, 189, 213
Maynard Smith, J., 15, 17, 235
Mbuti, 42–43, 195–198
McBurney, D. H., 237, 238
McCullough, M. E., 194
McGrath, A. E., 88–91, 120, 123
McIntyre, L. C., 66, 69, 70
McNeill, J. T., 105, 110, 123–124
McNeill, W. H., 155
Meeks, W. A., 150
Memes, 53

methodological individualism. *See* individualism
Mexico, 79
Michod, R. E., 17, 37, 79, 86, 235
Micle, S., 137
migratory birds, 26
military groups, 225
Miller, K. R., 239
Miller, R. R., 244
Miller, W. W., 66
Mishnah Torah, 142–143
Montenegro, 200–203
Monter, E. W., 111
morality, 10, 21–25, 44, 54, 58, 123–124, 141, 202, 223–225; cultural evolution of, 28–35, 119–120; innate psychological mechanisms, 25–28, 28–30, 119–122; religious expression of, 40–43, 122–123
Mormonism, 148
Mother Teresa, 80
multilevel selection, 10, 15–17, 20–25, 34–35, 43, 69–70, 83, 113, 119, 135, 138, 141, 175, 187, 203–204, 235, 239, 242, 243
mutual criticism, 58, 65

Naphy, W. G., 109
Nathan, H., 137
natural experiments, 72–73
natural selection, 7, 72. *See also* group selection; individual selection; multilevel selection
naturism, 52
Near, D., 244
nectar-robbing, 38
New Testament, 207–216
Nisbett, R. E., 79
nonadaptation hypotheses, 44–45, 86, 164, 215–216
norms, 21–25, 27
Nowak, M. A., 192
Nuer, 55, 56–61, 78–79, 106, 137, 199, 227, 240

Odling-Smee, J., 240, 244
Olson, J. E., 110

Oppenheim, A., 137
organismic concept of groups, 1, 3, 5, 6–
    7, 9, 17, 18, 20, 43, 66–68, 85, 86,
    95–96, 124, 125–126, 189, 236,
    244
Ostrer, H., 137
Ostrom, E., 27, 241

Pagels, E., 205–216, 218
Paley, W., 116
Pargament, K. I., 193
Pasteels, M., 33
Paul, 7, 212
Payne, J. W., 34
Peocilia reticulata, 71–73, 91–92, 188
Perrin, A., 111–112
pest control, 128–129
Peter, 211
Peterson, D., 38
phenotype-environment correlation, 72,
    240
phenotypic plasticity, 74
Phillips, K. P., 79
Pickering, W. S. F., 66
Pilate, Pontius, 212–214
plagues. See disease
Plotkin, H., 31
Pomiankowski, A., 40, 237
population genetics, 137
Power, C., 238
Prager, D., 136, 138, 139, 140–141
prayer, 103–104
prediction, 172–173, 189–195
Price, J., 30
production, 172, 173–177, 194–195
profane. See sacredness
Protestant Reformation, 88–91, 111–114,
    122, 140, 241
Protestant work ethic, 120
Provine, W. B., 238
proximate explanation, 67–68, 74–76,
    170–177, 188, 215, 241
psychological mechanisms, 25–28, 119–
    122, 189–195
public good, 19, 22
punen, 22–25
Putnam, R. D., 79, 161, 204

Quakers, 185, 209

Ragin, C. C., 222
Rapoport, A., 190, 191, 236
Rappaport, R., 130
rational choice theory, 25, 48–52, 79–85,
    131–132, 160, 169, 171, 173–174,
    187
rational thought, 40–43, 73–79, 122–
    123, 129, 227–230
Rauscher, F., 238
Rawls, J., 99
reciprocal altruism, 12, 22
Redd, A. J., 137
reductionism. See holism
Reid, J. K. S., 106
religion: definition of, 3, 220–230; evolu-
    tionary theories of, 44–45; in hunter-
    gatherer societies, 22–25, 42–43, 237;
    religious expression of morality, 40–
    43, 122–123; social scientific theories
    of, 47–85, 161–188; strictness of,
    163–164
resource conservation, 15
retaliation, 190, 194, 198, 199
revenge, 192
reverse dominance, 21
Richerson, P. J., 26, 31, 34, 40, 225, 226–
    227, 238, 244
ritual, 61–65
robber's cave experiment, 27
Rolls, E., 77, 84
Roman Empire, 147–157
Roozen, D. A., 163
Ruse, M., 239
Russell, J. C., 151

sacredness, 54, 55, 65, 200, 216, 225–
    227
Sahlins, M. D., 28
Sanachiora-Benerocetti, S., 137
Scher, S. J., 238
Schwartz, J., 243
science, 1, 6, 41, 44–46, 70–73, 83–85,
    87–88, 115–118, 124, 125–126, 132–
    133, 157–159, 161, 188, 189–230,
    239

Second Coming, 104–105
sects, 182
secular utility, 53–54, 86, 125–160, 161–188
Seeley, T. D., 20, 33–34, 235
self-interest, 2, 12, 64, 80, 166, 235, 240
selfish gene theory, 16, 18, 236–237
selflessness, 55, 175–177
Servetus, M., 114–115
sex ratio, 151–152
sharing, 21
Sherif, C. W., 27
Sherif, M., 27
Sigmund, K., 192
Simon, H. A., 26
Singer, I. B., vii, 144–146, 184
Skyrms, B., 16
Smith, H., 176
Sober, E., 12, 19, 20, 67, 116, 166–167, 170, 190, 223, 235, 237, 240, 243, 244
social control, 19–20, 24–25, 61, 97–98, 105–115, 127–128, 156–157, 211, 243
social dilemma experiments, 27
social identity theory, 27, 139
social insects, 20, 33–34
social norms. See norms
social sciences, 47–48, 78–79, 83–85, 161–162, 187–188
Somé, M. P., 63–65
Spandrel hypotheses. See byproduct hypotheses
Sperber, D., 238
Srinivasan, M., 26
St. Clair, U., 222
Standard Social Science Model (SSSM), 29
Stark, R., 41, 48–52, 55, 80, 82, 118, 131–132, 148–157, 179–181, 186–187, 220, 222, 229
Sterelny, K., 236, 238
Stevens, A., 30
structure. See communitas
subak, 127–128, 131
superstition, 22–25
Swanson, G. E., 241
Swenson, W., 239

symbolic systems, 54–55, 59, 225–227
symbolic thought, 26, 31
Szathmary, E., 17, 235

Tajfel, H., 139
Telushkin, J., 136, 138, 139, 140–141
Ten Commandments, 94, 96–97
Thoresen, C. E., 193
Thornhill, R., 231
Tit-for-Tat (TFT), 190–194, 236
Tocqueville, A. de, 27, 79, 198
Tomasello, M., 6, 238
Tonomi, G., 238
Tooby, J., 11, 26, 28–29, 85, 238
trait-groups, 15–16, 17, 131, 236
Trost, M. R., 174
Turelli, M., 239
Turnbull, C. M., 42–43, 195–198
Turner, V., 61–63, 224
Tylor, E. B., 41, 44

ultimate explanation, 67–68, 74–76, 170–177, 188, 241
unifying systems, 219–233
universal brotherhood, 10, 217–218
utility maximization, 80, 82

Van Schaik, C., 38
Viret, P., 89

Wade, M. J., 239
Walker, J. M., 27
Wallace, R. S., 105
warning cries, 8, 10, 12–13, 14–15, 235
water temple system of Bali, 126–133, 177–178, 241–242
Watkins, J. W. N., 66–67
Weber, B. H., 239
Weber, M., 120
Wegner, D. M., 7, 33
Wehner, R., 26
Weiner, J., 36, 72, 92
Wesley, J., 183
Wheeler, W. M., 20
White, B. J., 27
Wilken, R. L., 150
Williams, G. C., 10, 11, 16, 239

Wilson, D. S., 12, 15, 19, 20, 29, 34, 41,
    68, 74, 99–100, 110, 166–167, 170,
    190, 193, 223, 228, 229, 235, 237,
    238, 239, 240, 242–243, 244
Wilson, E. O., 26, 44, 238, 244
Wilson, L. S., 80
Wilson, M., 96
Wood, E. T., 137
Wrangham, R., 38

Wright, R., 237
Wright, S., 238–239
Wynne-Edwards, V. C., 15

Yoshimura, J., 74

Zande, 55
Zen Buddhism, 1